STABILITY ANALYSIS OF DISCRETE EVENT SYSTEMS

Adaptive and Learning Systems for Signal Processing, Communications, and Control

Editor: Simon Haykin

Werbos / THE ROOTS OF BACKPROPAGATION: From Ordered Derivatives to Neural Networks and Political Forecasting

Krstić, Kanellakopoulos, and Kokotović / NONLINEAR AND ADAPTIVE CONTROL DESIGN

Nikias and Shao / SIGNAL PROCESSING WITH ALPHA-STABLE DISTRIBUTIONS AND APPLICATIONS

Diamantaras and Kung / PRINCIPAL COMPONENT NEURAL NETWORKS: Theory and Applications

Tao and Kokotović / ADAPTIVE CONTROL OF SYSTEMS WITH ACTUATOR AND SENSOR NONLINEARITIES

Tsoukalas and Uhrig / FUZZY AND NEURAL APPROACHES IN ENGINEERING

Hrycej / NEUROCONTROL: Towards an Industrial Control Methodology

Beckerman / ADAPTIVE COOPERATIVE SYSTEMS

Cherkassky and Mulier / LEARNING FROM DATA: Concepts, Theory, and Methods

Passino and Burgess / STABILITY ANALYSIS OF DISCRETE EVENT SYSTEMS

STABILITY ANALYSIS OF DISCRETE EVENT SYSTEMS

Kevin M. Passino

Kevin L. Burgess

Department of Electrical Engineering
The Ohio State University

A Wiley-Interscience Publication
JOHN WILEY & SONS, INC.
New York / Chichester / Weinheim / Brisbane / Singapore / Toronto

About the Cover: The basin is a quadratic Lyapunov function. The lines that swirl to the bottom of the basin show many possible trajectories that can evolve in a discrete event system; hence, the figure illustrates a stable discrete event system since for this case the Lyapunov function is strictly decreasing along all the system trajectories.

Acquisitions Editor: Andrew Smith
Assistant Editor: Mary Lynn
Associate Managing Editor: Rosalyn Farkas

This text is printed on acid-free paper ♾.

Library of Congress Cataloging in Publication Data:

Passino, Kevin M.
Stability analysis of discrete event systems / Kevin M. Passino and Kevin L. Burgess.
p. cm. — (Adaptive and learning systems for signal processing, communications, and control)
"A Wiley-Interscience publication."
ISBN 0-471-24185-7 (cloth : alk. paper)
1. System analysis. 2. Discrete-time systems. 3. Stability. I. Burgess, Kevin L., 1968– . II. Title. III. Series.
QA402.P325 1998
003'.85—dc21 97-26396

Printed in the United States of America
10 9 8 7 6 5 4 3 2 1

To my parents, Stan and Dolores (K. M. P.)

To Erin (K. L. B.)

Contents

Foreword xi

Preface xiii

1 Introduction 1

1.1 Emergence of Discrete Event Systems 1

1.2 Need for Analysis 2

1.3 What This Book is About 3

1.4 Summary 4

1.5 For Further Study 5

1.6 Problems 5

2 Modeling 7

2.1 Overview 7

2.2 A Discrete Event System Model 9

2.3 Examples 10

2.3.1 Single Buffer Machine 11

2.3.2 Load Balancing System 12

2.4 Petri Net Model 14

2.5 Petri Net Examples 16

2.5.1 Production Network 16

2.5.2 Network of Computers 18
2.6 Summary 18
2.7 For Further Study 19
2.8 Problems 19

3 Stability Concepts and Analysis Techniques 23
3.1 Overview 23
3.2 Mathematical Preliminaries 24
3.3 Stability Concepts 25
3.4 Stability Analysis Techniques 28
3.5 Reachability and Cyclic Properties 34
3.6 DES Applications 35
3.6.1 Automata and Finite State Systems 35
3.6.2 Petri Nets 36
3.6.3 Rate Synchronized Manufacturing Line 38
3.7 Summary 39
3.8 For Further Study 39
3.9 Problems 41

4 Load Balancing in Computer Networks 45
4.1 Overview 45
4.2 A Load Balancing Problem without Delays 46
4.2.1 Asymptotic Convergence to a Balanced State 49
4.2.2 Exponential Convergence to a Balanced State 51
4.3 Load Balancing Problem Generalizations 57
4.3.1 Generalized Load Passing Conditions 57
4.3.2 Virtual Load 58
4.3.3 Discrete Load 60
4.4 The Load Balancing Problem with Delays 64
4.4.1 Asymptotic Convergence to a Balanced State 68
4.4.2 Exponential Convergence to a Balanced State 74
4.5 Summary 81
4.6 For Further Study 81
4.7 Problems 82

5 Scheduling in Manufacturing Systems **85**

5.1 Overview 85

5.2 FMS Scheduling: A Distributed Approach 86

5.2.1 Machine Model and Stream Constraints 86

5.2.2 Clear-A-Fraction Policy 89

5.2.3 Clear-Average-Oldest-Buffer Policy 92

5.2.4 Random Part Selection Policy 95

5.2.5 Characterization of the Output Stream 98

5.2.6 Network Element Analysis 98

5.3 Stability of Distributed FMS 106

5.3.1 Modeling the FMS 106

5.3.2 Designing Stable Distributed FMS 108

5.3.3 Computing the Buffer Bounds 112

5.3.4 Artificial Implementation of the Stream Modifiers 113

5.3.5 Buffer Bounds: Theoretical and Simulated 113

5.4 FMS Scheduling: A Centralized Approach 114

5.4.1 FMS Description and Notation 115

5.4.2 Global Synchronous Clear-A-Fraction Policy 117

5.4.3 Global Synchronous Periodic Clearing Policy 123

5.5 Scheduling Examples: Centralized vs. Distributed 130

5.5.1 Highly Re-entrant Line 131

5.5.2 Feedforward Line 133

5.5.3 Cellular Structure I 134

5.5.4 Cellular Structure II 138

5.5.5 Alternatives to GSCLB 140

5.6 Summary 142

5.7 For Further Study 142

5.8 Problems 143

6 Intelligent Control Systems **147**

6.1 Overview 147

6.2 Expert Control Systems 148

6.2.1 The Expert Controller 149

6.2.2 The Expert Control System: Model and Approaches to Analysis 160

6.2.3 Expert System Verification 165

6.2.4 *Expert Control of a Surge Tank* 166
6.2.5 *Expert Control of a Flexible Manufacturing System* 172
6.3 *Planning Systems for Control* 180
6.3.1 *Planners as Controllers* 180
6.3.2 *Analysis of Closed-Loop Planning Systems* 181
6.4 *Intelligent and Autonomous Control* 182
6.5 *Hybrid System Theory and Analysis* 187
6.6 *Summary* 188
6.7 *For Further Study* 188

References **191**

Index **201**

Foreword

The demands of modern technology have resulted in the synthesis and implementation of systems of increasing complexity. Such *contemporary systems* frequently defy simple and tidy descriptions by classical equations of the type encountered in science and applied mathematics (such as, for example, ordinary differential and ordinary difference equations, partial differential equations, functional differential equations, Voltaire integrodifferential equations, and the like). Examples of such contemporary systems include *hybrid systems*, which are described by a mixture of equations of the type enumerated above, where different system components may evolve along different kinds of "time." In their most complete descriptions, parts of such systems may involve "equation free" characterizations (such as Petri nets and logic statements) of *discrete events*, or, more generally, the entire system may be viewed as a *discrete event system*.

Initially, the study of the majority of discrete event systems seemed to employ ad hoc procedures and simulation-based methods. However, because of their complexity, more recent analysis and synthesis techniques for such dynamical systems appear to rely increasingly on qualitative techniques, which have their roots in well-grounded classical methodologies. At the forefront of these have been stability investigations by direct methods (particularly Lyapunov stability of invariant sets (such as equilibria) and the boundedness of motions (Lagrange stability) of dynamical systems).

The present work presents, perhaps for the first time, a systematic qualitative treatment of discrete event dynamical systems. This is accomplished by first providing a foundation (modeling issues, followed by general stability aspects) and then by applying this foundation to problems that arise in three important classes of systems: load balancing in computer networks, scheduling in manufacturing systems, and intelligent control systems.

The authors of this monograph have made several of the initial contributions in this area of dynamical systems. This has enabled them to produce a highly original and insightful piece of work on this subject, which reads rather well.

ANTHONY N. MICHEL

Notre Dame, Indiana, 1997

Preface

Discrete event systems (DES) are dynamical systems, which evolve in time by the occurrence of events, at possibly irregular time intervals. Examples of DES include manufacturing systems, computer networks, traffic systems, and many others. Stability theory plays a central role in the proper operation of a dynamical system. For instance, we may need to know that the part queues in a manufacturing system will remain bounded so that they do not overflow or create large inventories, or that the buffers holding tasks in a computer network equalize under certain conditions to avoid under-utilization of computing resources. This book focuses on the characterization and analysis of such stability properties using a Lyapunov approach. Basically, we use a system-theoretic methodology for the modeling, design, and verification of the correct operation of DES. Verification of the correction operation of DES is clearly an important issue to study, especially considering their ubiquitous presence in society today and our heavy dependence on their reliable operation.

After a brief introduction to DES and the focus of this book in Chapter 1, in Chapter 2 we present two different models for DES. In Chapter 3 we explain how to characterize and analyze stability properties of DES. In Chapters 4–6 we conduct extensive case studies in the use of the modeling and analysis methods of Chapters 2 and 3. In particular, in Chapter 4 we study load balancing problems in computer networks, and in Chapter 5 we study scheduling of manufacturing systems. In Chapter 6 we study stability of intelligent control systems that can be modeled with DES models.

Each chapter has an overview, a summary of the main topics that have been covered, and a section "For Further Study," which indicates what literature one would study to pursue topics at a deeper level. There are many examples used throughout the text; indeed Chapters 4–6 are extended examples ("case studies"). Chapters 1–5 have homework problems. These include typical homework problems that can be used to strengthen the student's understanding of the material in the chapter and problems that have a star ("$\star$") next to them. In such starred problems we provide only a general description of the problem and let the reader fill in the details of the problem specification and its solution (i.e., they take on more of a research character since they are open-ended).

Portions of this text have been used in a graduate level course on discrete event systems. The book could be used as a textbook for such a course or in a role where the instructor would also cover some of their own topics, in addition to the ones covered here. The book is reasonably self-contained; however, the student with a course in discrete-time systems, automata, or nonlinear control (particularly stability theory) would probably find it more user-friendly. The book will also be useful to researchers in the field of DES and hybrid systems, since it provides the major foundations for a stability theory for DES and it provides some challenging DES applications.

We would like to acknowledge the help from Anthony N. Michel, Alfonsus D. Lunardhi, and Panos J. Antsaklis, who worked with us on some of the research that this book is based on. The appropriate references to the material that we co-authored with them are contained in the Bibliography. We would like to thank Veysel Gazi for proofreading a portion of the manuscript. We would like to acknowledge the financial support of the Engineering Foundation and National Science Foundation grants IRI-9210332 and EEC-9315257. We thank Andrew Smith, Mary Lynn, Rosalyn Farkas and the whole production and editing team at John Wiley and Sons for all their help. Finally, we would thank our wives, Annie and Erin, for providing a supportive environment for us while we were writing this book.

KEVIN PASSINO AND KEVIN BURGESS

Columbus, Ohio, November 1997

1
Introduction

1.1 EMERGENCE OF DISCRETE EVENT SYSTEMS

Discrete event systems (DES) are dynamical systems, which evolve in time by the occurrence of events at possibly irregular time intervals. Sometimes the term "discrete event dynamical systems" (DEDS) is used to emphasize that DES are dynamical systems. Examples of DES include the following:

- Manufacturing systems (e.g., machines, sequencing operations, inspection operations, etc.).
- Robotic systems (e.g., pick and place operations, assembly operations, etc.).
- Computers and computer networks (e.g., logic circuits, interfaces, communication protocols, etc.).
- Communication networks (e.g., asynchronous transfer mode (ATM) networks).
- Traffic systems (e.g., traffic signals, ramp metering, automated highway systems' platooning operations, etc.).
- Batch operations in the process industries (e.g., sequencing of mixing operations or sequencing of processing operations).
- General nonlinear difference equations (e.g., expert control systems or planning systems as controllers with a discrete event plant).

DES tend to be systems that have been invented, developed, and constructed by scientists and engineers and, in this sense, they do not tend to occur "naturally" as some traditional dynamical systems do. Many DES are then relatively new systems, which often have significant complexity, and there are likely to be more DES that are invented in the future.

Many DES are distributed in the sense that their components may be physically separated. For instance, in a manufacturing system there may be several machines with a robotic transporter in between them. In a computer system there may be computers distributed all over the world and the computer network that interconnects them may act as a DES. In a traffic system we have many interconnected intersections, each of which we may view as a local DES. Other DES are centralized. For instance, a logic circuit may be physically located in one place and we may be interested in studying only its local properties and not those of the systems that it interconnects with.

1.2 NEED FOR ANALYSIS

The emergence of complex DES such as those mentioned in the previous section has dictated the need to try to fully understand their behavior and to show that they operate properly and in a reliable fashion. Moreover, it has dictated the need for design methodologies for their construction. Currently, some DES are constructed and verified with software engineering methodologies (e.g., the "waterfall" or "spiral" methods), but many are constructed in an ad hoc fashion using "seat-of-the-pants" engineering methods. This is an unfortunate situation as such ad hoc methodologies can lead to costly mistakes or inefficient or incorrect operation of a DES.

In this book we take a more formal approach to the analysis of properties of DES, where we begin with a formal mathematical model of the DES. Such a model can provide many benefits, including a clear understanding of the following:

- The operation of the DES.
- Whether or not it will meet some specifications (e.g., whether it will operate fast enough to achieve some task).
- Ideas on how to redesign the DES so that it will properly meet specifications.

Given a mathematical model of a DES we can also study several properties of the DES that will allow us to determine if it is operating according to the specifications. For instance, we may want to study the following properties of systems to make sure that they are operating properly:

- *Manufacturing Systems.* Boundedness properties of queues or whether a queue will eventually have a buffer level less than some bound so that we can implement a fixed length queue.

- *Robotic Systems.* Whether the sequence of operations in a pick and place task will succeed even if there are certain disturbances or whether an assembly/inspection operation will succeed.
- *Computers and Computer Networks.* Will the logic circuit perform properly under all possible sequences of inputs, whether a computer communication protocol will ensure safe and reliable transmission of information between computers, or whether we can design a system that will seek to balance the processing load between a set of computers.
- *Traffic Systems.* Boundedness properties for the queues and how to switch the traffic lights so that the queues from one intersection will not interfere with another intersection, whether the ramp metering strategy will maintain stable and efficient traffic flow on the main highway, or whether the platooning operations in an automated highway system will ensure that passing, merging, exiting, and entering a highway are all done in a safe fashion.

Clearly there are many properties of DES that need to be studied, and that we need to verify, before such a system can be put into operation.

1.3 WHAT THIS BOOK IS ABOUT

In this section we will overview the contents of this book in more detail than in the preface and will state the basic objectives of the book.

In Chapter 2 we will introduce a general DES model and give a manufacturing system example and a computer network load balancing example to illustrate how to use the model. In addition, we introduce a variety of Petri net models and provide a production network and computer network example to illustrate those modeling methodologies.

In Chapter 3 we provide mathematical definitions of all the stability properties that we will study in this book. These include stability in the sense of Lyapunov, asymptotic stability, exponential stability, Lagrange stability, uniform boundedness, and uniform ultimate boundedness. Following this we provide sufficient conditions for a DES to possess such properties in terms of the existence of a specified "Lyapunov function" that satisfies certain properties. Finally, we explain how the Lyapunov framework applies to automata, finite state systems, and Petri nets, and provide a simple "rate synchronized manufacturing line" as an illustrative example. The Petri net studies are used to illustrate uniform boundedness and uniform ultimate boundedness.

In Chapter 4 we present the first of our three case studies in stability analysis of discrete event systems. This case study involves studying the stability properties of a variety of load balancing problems in computer networks. These include load balancing problems with and without sensing and load transfer delays, discrete and continuous load, and virtual load. The stability properties that are studied include stability in the sense of Lyapunov, asymptotic stability, asymptotic stability in the large, exponential stability, and exponential stability in the large.

In Chapter 5 we present our second case study in stability analysis. This case study involves the development of scheduling policies for manufacturing systems that result in bounded buffers. In particular, we show that several scheduling policies for an isolated machine result in bounded buffers. In addition, we show that certain network elements such as a multiplexer, demultiplexer, stream modifier, and bounded delay can be implemented with bounded buffers. Next, we show how to use the stability results on the isolated network elements to form a stable flexible manufacturing system (i.e., we show how to implement a distributed scheduler). In the second part of the case study we study centralized schedulers that use information on part flow paths in a flexible manufacturing system to make scheduling decisions. We study several policies and then we compare the performance of the centralized vs. distributed schedulers. Overall, the stability property that is studied in Chapter 5 is Lagrange stability.

In Chapter 6 we present our third and final case study in stability analysis of DES. This case study involves the analysis of stability properties (e.g., asymptotic stability in the large) of expert control systems and some discussions on analysis of properties of control systems that use planning systems as controllers. Moreover, we provide a brief overview of the area of intelligent autonomous control and explain the need for hybrid system analysis for such systems.

Overall, the three goals of this book are the following:

1. To provide several modeling approaches for DES.
2. To provide several stability properties and methods to verify that a system possesses these properties.
3. To show how stability analysis is conducted for complex applications (and not just simple academic examples).

While Chapter 3 is somewhat theoretical, overall, the book is quite accessible once Chapters 2 and 3 are understood. It is our hope that the wealth of examples and case studies that are used in the text will help you to bridge the gap between theory and practice of using stability analysis for DES.

1.4 SUMMARY

In this chapter we have explained how DES are emerging as an important class of dynamical systems. We have explained the importance of analysis and have summarized each of the chapters in the text. The topics covered in this chapter include the following:

- A list of examples of DES.

- How DES can be distributed or centralized.
- Why it is important to perform modeling of DES.
- Why it is important to analyze and verify DES properties, particularly stability.

This chapter simply provides some motivational material for the text and overviews the chapters. The later chapters depend on it in no significant way.

1.5 FOR FURTHER STUDY

While the "For Further Study" sections in all the later chapters will provide a wealth of references, here we simply provide an overview of some general references on DES. These include the books [19, 44, 66, 37] and the special issue [89]. Others are appearing regularly and it is recommended that the reader perform a literature search to obtain more recent titles.

For instance, for more recent work in all areas of DES theory see the proceedings of the IEEE Conference on Decision and Control, the American Control Conference, the European Control Conference, and the International Federation on Automatic Control World Congress. Major journals to keep an eye on include the *IEEE Trans. on Automatic Control*, *Discrete Event Dynamic Systems: Theory and Applications*, *IEEE Trans. on Control Systems Technology*, *IEEE Control Systems Magazine*, *Systems and Control Letters*, *Automatica*, *Control Engineering Practice*, *International Journal of Control*, SIAM journals, ACM journals, and several others.

Recently, a relatively complete treatment of the field of control has appeared in [45] and the reader may want to use this for finding additional studies on modeling and stability analysis in conventional control.

1.6 PROBLEMS

Problem 1.1 (Discrete Event System Applications):

(a) Compile a list of discrete event systems from the area of computers and computer networks.

(b) Compile a list of discrete event systems from the area of manufacturing systems and robotics.

Problem 1.2 (Discrete Event System Analysis):

(a) Define what "reachability" or "controllability" would mean for a discrete event system and give examples of DES that you would like to possess such properties.

(b) Define what "detectability" or "observability" would mean for a discrete event system and give examples of DES that you would like to possess such properties.

(c) Why would it be important to analyze the real-time properties of DES (e.g., that a DES could complete a task within a given time frame). Give specific examples of DES to support your explanation.

Problem 1.3 (Discrete Event System Literature Search):

(a) Do a literature search on the field of discrete event systems that focuses on finding all the books written on this topic to date.

(b) Perform a literature search on the specific area of stability analysis of discrete event systems. To do this you should consider not only the books you found in (a) but also the journals and conferences listed in Section 1.5. You may also consider using the Bibliography at the end of this text.

2

Modeling

2.1 OVERVIEW

Discrete event systems (DES) are dynamical systems which evolve in time by the occurrence of events at possibly irregular time intervals. Some examples include manufacturing systems, computer networks, logic circuits, and traffic systems. For a manufacturing system suppose that you have a single machine that can process parts one at a time and suppose that it can be reconfigured to process different part types. Suppose that there are buffers (queues) where different part types wait to be processed, and that there is a scheduling "policy" that selects which part type to process at each time instant. There could be a "set-up" time (a delay) that occurs due to the time it takes to reconfigure the machine to process one part type when it is currently processing a different part type (e.g., there may be a need to change drill bits). "Events" in such a system would be the arrival of a part at a buffer, commencement of the processing of a part, or finishing the processing of a part. Often the parts will not arrive at discrete synchronous time instants (e.g., once every second) but "asynchronously" at any random time. Such a system is a dynamical system and models can be used to represent it and the other discrete event systems mentioned above.

In this chapter we will introduce two discrete event system models. The first one is a nonlinear difference equation that allows for modeling nondeterministic behavior in the sense that if the system is at any state there is a *set* of events that can occur, and depending on which one occurs the system can transition to different states. For instance, for the machine scheduling example above the system can be at one state and the state could asynchronously transition to many different values, depending on what part types arrive in the buffers, or when the part that is being processed is completed. We use a simple manufacturing system and load balancing system to illustrate the modeling process for the discrete event system model.

The second model that we introduce is the Petri net model. After introducing the notation and terminology of the general and extended Petri net models we use production network and computer network examples to illustrate the modeling approach. The Petri net models that we introduce are actually a special case of the DES model in Section 2.2, in the sense that any system that the Petri net can represent, the one in Section 2.2 can also. The converse does not hold. That is, one can easily specify a manufacturing system that, for example, the general Petri net model cannot represent. Due to this fact, we will ultimately find that Petri net models will be of only somewhat limited use in this book, even though they often provide for a very intuitive modeling methodology.

We will use the models introduced in this chapter throughout the remainder of this book. We will also introduce a variety of other DES models throughout the book. In particular:

- In Chapter 3 we will introduce an automata model.

- In Chapter 4 we will use the nonlinear difference equation model of this chapter for representing a computer network load balancing problem.

- In Chapter 5 we will produce a model of a flexible manufacturing system that does not adhere to the models in this chapter since it is easier to model such a system with a formalism that is more closely tied to that application.

- In Chapter 6 we will introduce a slightly generalized version of the nonlinear difference equation model in this chapter to represent an expert control system. Moreover, in Chapter 6 we will briefly discuss hybrid system models.

Upon completion of this chapter the reader will have a basic understanding of how to model a discrete event system. It is important that the reader have an open mind about which DES model should be used. There does not seem to be a universally accepted model or models (like there are for linear systems) and, hence, it is important that the reader understand many modeling formalisms and be able to invent some that might be particularly useful for the application at hand. All the remaining chapters in this book depend on the reader's understanding of this one. However, the reader not interested in Petri nets can skip that material here and in Chapter 3 since the remainder of the book does not depend on it in any way.

2.2 A DISCRETE EVENT SYSTEM MODEL

We will study the stability of systems that can be accurately modeled with

$$G = (\mathcal{X},\ \mathcal{E},\ f_e,\ g,\ \boldsymbol{E}_v). \tag{2.1}$$

Here, $\mathcal{X}$ is the set of states. In this book we will sometimes use the term "state" somewhat loosely. At times $x \in \mathcal{X}$ will represent the state of the system in the traditional Markovian sense. Other times we will refer to x as a state, but it may not adhere to the traditional definition. We use $\mathcal{E}$ to denote the set of events. State transitions are defined by the operators

$$f_e : \mathcal{X} \longrightarrow \mathcal{X}, \tag{2.2}$$

where $e \in \mathcal{E}$. An event, e, may only occur if it is in the set defined by the enable function,

$$g : \mathcal{X} \longrightarrow \mathcal{P}(\mathcal{E}) - \{\emptyset\}, \tag{2.3}$$

where $\mathcal{P}(\mathcal{E})$ denotes the power set of $\mathcal{E}$. We only require that f_e be defined when $e \in g(\boldsymbol{x})$. Notice that according to the definition of g, it can never be the case that no event is enabled. We can, however, model "deadlock" (the case where there is no enabled event) by defining a "null event," e^0, so that

$$f_{e^0}(\bar{\boldsymbol{x}}) = \bar{\boldsymbol{x}}, \tag{2.4}$$

where $\bar{\boldsymbol{x}} \in \mathcal{X}$ is any state at which the system is deadlocked.

We associate "logical time" indices with the states and events so that $\boldsymbol{x}_k \in \mathcal{X}$ represents the state at time $k \in \{0, 1, 2, \ldots\} = \mathbb{N}$ (the set of natural numbers) and $e_k \in g(\boldsymbol{x}_k)$ represents an *enabled* event at time $k \in \mathbb{N}$. Notice that there can be just one state at time k, but that many events may be enabled at time k. Should an enabled event e_k occur, then the next state, $\boldsymbol{x}_{k+1}$ is defined by

$$\boldsymbol{x}_{k+1} = f_{e_k}(\boldsymbol{x}_k). \tag{2.5}$$

We now define "state trajectories" and "event trajectories." A state trajectory is any sequence $\{\boldsymbol{x}_k\} \in \mathcal{X}^{\mathbb{N}}$ (the set of state sequences) such that $\boldsymbol{x}_{k+1} = f_{e_k}(\boldsymbol{x}_k)$ for some $e_k \in g(\boldsymbol{x}_k)$ for all $k \in \mathbb{N}$. An event trajectory is any sequence $\{e_k\} \in \mathcal{E}^{\mathbb{N}}$ (the set of event sequences) such that there exists a state trajectory, $\{\boldsymbol{x}_k\} \in \mathcal{X}^{\mathbb{N}}$, where for every $k \in \mathbb{N}$, $e_k \in g(\boldsymbol{x}_k)$. The set of all such event trajectories is denoted by $\boldsymbol{E} \subset \mathcal{E}^{\mathbb{N}}$. Notice that corresponding to a given event trajectory, there can be only one state trajectory. In general, however, an event trajectory that produces a given state trajectory is not unique. Notice that all state and event trajectories must be infinite sequences.

Let $\boldsymbol{E}_v \subset \boldsymbol{E}$ denote a set of what we call "valid" event trajectories that we assume is specified as part of the modeling process. Let $\boldsymbol{E}_v(\boldsymbol{x}_0)$ be the set of valid event trajectories when the initial state is $\boldsymbol{x}_0 \in \mathcal{X}$. The framework provides another mechanism for further pruning $\boldsymbol{E}$. $\boldsymbol{E}_a \subset \boldsymbol{E}_v$ is the set of what we call "allowed" event trajectories. Including $\boldsymbol{E}_a$ in our model yields a great deal of modeling power. For example, we will make use of $\boldsymbol{E}_a$ to model the decision-making policies that we impose on our systems in Chapter 4.

If we fix $k \in \mathbb{N}$, then E_k denotes the sequence of events $e_0, e_1, \ldots, e_{k-1}$, and the $E_k E \in \boldsymbol{E}_v(\boldsymbol{x}_0)$ is used to denote the concatenation of E_k with a sequence of infinite length $E = e_k, e_{k+1}, \ldots$ such that $E_k E \in \boldsymbol{E}_v$ ($E_0 = \emptyset$, the string with no elements in it which we also use to denote the empty set). If E is a string then $|E|$ denotes the length of the string (i.e., the number of elements in the string). Let

$$\boldsymbol{E}_v^f = \{E' : E'E \in \boldsymbol{E}_v, |E'| < \infty\}$$

(i.e., the set of all finite length valid event trajectories). Let

$$X : \mathcal{X} \times \boldsymbol{E}_v^f \times \mathbb{N} \to \mathcal{X}.$$

The value of the function $X(\boldsymbol{x}_0, E_k, k)$ will be used to denote the state reached at time k from $\boldsymbol{x}_0 \in \mathcal{X}$ by application of event sequence E_k such that $E_k E \in \boldsymbol{E}_v(\boldsymbol{x}_0)$. For fixed $\boldsymbol{x}_0$, the functions $X(\boldsymbol{x}_0, E_k, k)$, where $E_k E \in \boldsymbol{E}_v(\boldsymbol{x}_0)$, are called "motions."

We will often assume that it is possible to define a metric on the set of states and hence obtain a metric space for the study of stability. This assumption will become clear in Chapter 3 when we define stability concepts and the methods that we use to analyze them. When we study the applications in Chapters 4, 5, and 6 we will see that assuming we can form a such a metric space is not overly restrictive, especially if one considers real-world applications, where it is often possible to define a distance measure.

It is possible to define a graph that represents the above model. To do this we make the states nodes (circles) and label the arcs (arrows) with the events. For simple systems the graph representation of the model can provide good intuition into the dynamics of the system. For realistic real-world applications such a graph can become unwieldy as its size can be very large, since there can easily be many states and events. Moreover, the graph representation is only a partial representation of the dynamics of the model G, since it will not in general represent the constraints imposed by the set of valid event trajectories $\boldsymbol{E}_v$. In the next section we provide a simple example of how to use the above model and will show how to use a graph to represent the model.

2.3 EXAMPLES

In this section we will provide two examples of how to model systems using the DES model G. In the first one we will study a single buffer machine, where we will also discuss how to construct the graph representation of the plant. In the second example we will study a computer network load balancing problem (a very simple version of the problem that we will study in detail in Chapter 4).

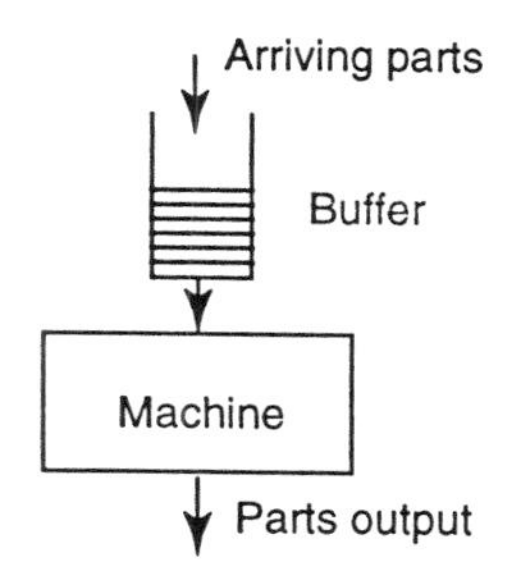

Fig. 2.1 Machine with buffer.

2.3.1 Single Buffer Machine

Suppose that we have a machine with a single buffer that can only process one part at a time. It takes some fixed, unknown, but bounded amount of time to process a part and when processing is completed, the parts are output from the system, as shown in Figure 2.1. Parts may arrive to be stored in the buffer at any finite rate. We will assume a part remains in the buffer even though the machine starts processing it.

We will use $\boldsymbol{x} = x$ to denote the buffer level and it will be the state of the system. The set of states is

$$\mathcal{X} = \{0, 1, 2, 3, \ldots\}. \tag{2.6}$$

Suppose that the event set is $\mathcal{E} = \{e^1, e^2, e^3\}$, where

$$\begin{aligned} e^1 &= \text{"a part arrives"} \\ e^2 &= \text{"the machine has finished processing a part"} \\ e^3 &= \text{"the machine has finished processing a part } \textit{and} \\ &\quad\ \text{a part arrives at the same time."} \end{aligned}$$

Notice that we use event e^3 to represent the possibility that two other events can occur at the same time. The enable function is defined by

$$g(x) = \{\{e^1\}, \{e^2\}, \{e^3\}\} \tag{2.7}$$

if $x > 0$, and

$$g(x) = \{\{e^1\}, \{e^3\}\} \tag{2.8}$$

if $x = 0$ (hence for $x = 0$ it takes zero time to process a part). Hence we see that for the state $x = 0$ the event e^2 is not enabled and this represents the fact that no parts are in the machine to be processed so there is no way that the machine can process a part. The state transition function is defined by

$$
\begin{aligned}
f_{e^1}(x) &= x + 1 \\
f_{e^2}(x) &= x - 1 \\
f_{e^3}(x) &= x.
\end{aligned}
$$

We will let $\boldsymbol{E}_v$ be the set of valid event trajectories. We consider $\boldsymbol{E}_v$ to constrain the types of event trajectories that can be generated by the model that is defined by $\mathcal{X}$, $\mathcal{E}$, f_e and g. The least restrictive way to define $\boldsymbol{E}_v$ is to let it contain any possible event trajectory that can be generated by the model that is defined by $\mathcal{X}$, $\mathcal{E}$, f_e and g. In this case the event trajectories are simply those that can evolve from the dynamical model. Let us call this set of valid event trajectories $\boldsymbol{E}$ as we had discussed in the previous section. There are many other ways to define $\boldsymbol{E}_v$ by pruning trajectories from $\boldsymbol{E}$. For instance, it could be the case that the number of parts that are processed is, on average, greater than or equal to the number of parts that arrive (so that parts will not, on average, accumulate in the buffer). For some other machine we may know that the event e^3 never occurs, e^1 occurs first, then the two events e^1 and e^2 alternate and this can easily be represented by letting $\boldsymbol{E}_v = \{\{e^1, e^2, e^1, e^2, \ldots\}\}$ (i.e., a set with one event trajectory in it). In this case we know that $\boldsymbol{E}_v$ only has one possible trajectory while in the other cases mentioned above there are an infinite number of possible trajectories. Clearly $\boldsymbol{E}_v$ adds significant representation power to our model.

Next, consider how we can use a graph to represent the dynamics that are modeled by $\mathcal{X}$, $\mathcal{E}$, f_e and g. Recall that the graph will not necessarily represent the constraints imposed by $\boldsymbol{E}_v$. A graph representation for the machine is given in Figure 2.2. The circles represent the states and the events label the arrows which represent the state transitions. If $\boldsymbol{E}_v = \boldsymbol{E}$, that is, all possible event trajectories, then this is inherently represented in the graph. However, if we prune event trajectories from $\boldsymbol{E}$, then the graph representation may not represent the entire model (e.g., if $\boldsymbol{E}_v = \{\{e^1, e^1, e^2, e^2, e^1, e^1, e^2, e^2, e^1, e^1, e^2, e^2, \ldots\}\}$ so that it will only cycle around the graph in a certain way) or may contain state transitions that are not even possible (e.g., if $\boldsymbol{E}_v = \{\{e^1, e^2, e^1, e^2, \ldots\}\}$ then only some of the arrows in Figure 2.2 are needed). In Chapter 5 we will consider much more complex machines and manufacturing systems.

2.3.2 Load Balancing System

Suppose that we have two computers connected on a network, each of which has the task of processing identical jobs (i.e., parallel processing). The jobs are stored in buffers and each computer can sense the other computer's buffer level and can pass jobs to the other computer (see Figure 2.3). There is a load balancing mechanism on each computer that seeks to balance (equalize)

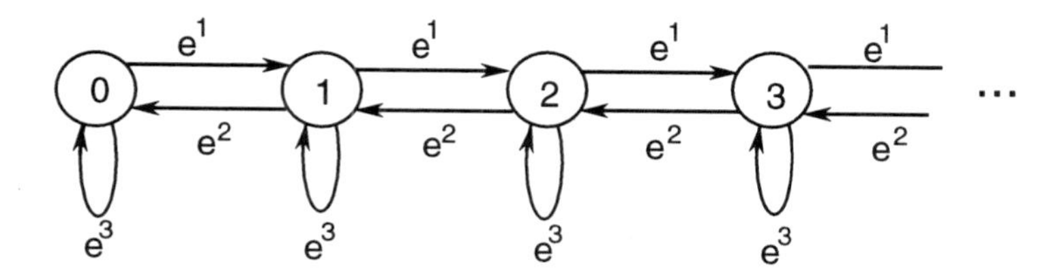

Fig. 2.2 Graph representation of the single buffer machine.

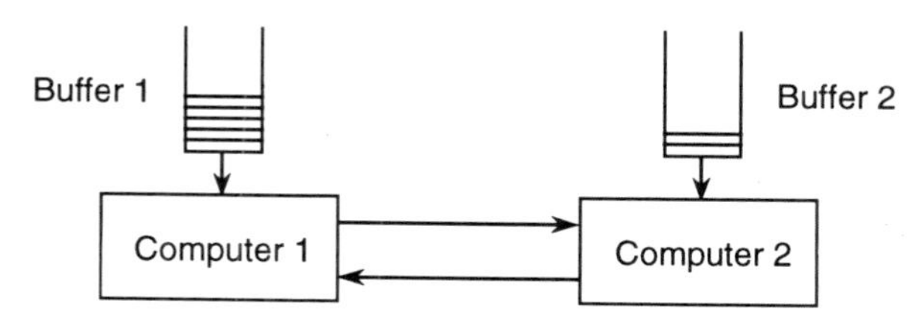

Fig. 2.3 Simple two-computer load balancing system.

the load between the two computers to try to ensure that at no time will one computer be overloaded while the other computer has no jobs to process.

The set of states is $\mathcal{X} = \mathbb{N}^2$ (i.e., 2×1 vectors of natural numbers) so that, for example, $\boldsymbol{x} = [x^1, x^2]^\top = [1, 5]^\top$ represents Computer 1 having one job and Computer 2 having five jobs (note that we use superscripts for the x components since we are using subscripts for the time index). The set of events $\mathcal{E}$ is given by

$$\mathcal{E} = \{e_{12}^\alpha, e_{21}^\alpha, e^0\},$$

where

$$\begin{aligned} e_{12}^\alpha &= \text{"pass } \alpha \in \mathbb{N} \text{ jobs from Computer 1 to Computer 2"} \\ e_{21}^\alpha &= \text{"pass } \alpha \in \mathbb{N} \text{ jobs from Computer 2 to Computer 1"} \\ e^0 &= \text{"pass no jobs" (a null event).} \end{aligned}$$

The enable function

$$g(x) = \begin{cases} \{\{e_{12}^\alpha\}\} & \text{if } x^1 > x^2 \text{ and } \alpha = \frac{1}{2}|x^1 - x^2| \\ \{\{e_{21}^\alpha\}\} & \text{if } x^2 > x^1 \text{ and } \alpha = \frac{1}{2}|x^1 - x^2| \\ \{\{e^0\}\} & \text{otherwise.} \end{cases}$$

The state transition function

$$\begin{bmatrix} x_{k+1}^1 \\ x_{k+1}^2 \end{bmatrix} = f_{e_{12}^\alpha}(x_k) = \begin{bmatrix} x_k^1 - \alpha \\ x_k^2 + \alpha \end{bmatrix}$$

$$\begin{bmatrix} x_{k+1}^1 \\ x_{k+1}^2 \end{bmatrix} = f_{e_{21}^\alpha}(x_k) = \begin{bmatrix} x_k^1 + \alpha \\ x_k^2 - \alpha \end{bmatrix}$$

$$\begin{bmatrix} x_{k+1}^1 \\ x_{k+1}^2 \end{bmatrix} = f_{e_{12}^\alpha}(x_k) = \begin{bmatrix} x_k^1 \\ x_k^2 \end{bmatrix}.$$

We let $\boldsymbol{E}_v = \boldsymbol{E}$ so that all event sequences defined by $\mathcal{X}$, $\mathcal{E}$, f_e, and g are possible. The load balancing policy (a type of controller) is inherently represented by f_e and g. Notice that the graph to represent the dynamics of the load balancing system can become very complex, especially if we attach additional computers to the load balancing network. For some initial load distribution the load will be transferred asynchronously between the computers to try to achieve a balanced load. If there are an odd number of jobs then clearly perfect balancing is not achievable. Also, notice that the model inherently assumes that no new jobs arrive or are processed by the computers. In Chapter 4 we will consider much more complex load balancing problems.

2.4 PETRI NET MODEL

Next, we introduce some Petri net models which are a special case of the model G in the sense that any system that the Petri nets can represent, the G model can also represent. It is for this reason that in Chapter 3 we only need to develop our stability theory for the model G, since the theory will automatically include systems that can be represented by Petri nets.

For our discussions on Petri nets we will adhere to the greatest extent possible to the somewhat standard notation in [61], where a Petri net

$$PN = (P, T, F, W, M_0),$$

where

- $P = \{p_1, p_2, \ldots, p_m\}$ is a finite set of *places* (represented with circles);
- $T = \{t_1, t_2, \ldots, t_n\}$ is a finite set of *transitions* (represented with line segments);
- $F \subset (P \times T) \bigcup (T \times P)$ is a set of arcs (represented with arrows);
- $W : F \to \{1, 2, 3, \ldots\}$ is an *arc weight function* (represented with numbers labeling arcs and assume for convenience that if $(p, t) \notin F$ or if $(t, p) \notin F$ we will extend the arc weight function so that $W(t, p) = W(p, t) = 0$ for these cases and the arrow will be omitted); and
- $M_0 : P \to \mathbb{N}$ is a (initial) *marking* (represented with dark dots, i.e., *tokens*, in places).

It is the case that $P \bigcap T = \emptyset$ and $P \bigcup T \neq \emptyset$. The Petri net structure is $\mathcal{N} = (P, T, F, W)$ so $PN = (\mathcal{N}, M_0)$. The Petri net PN is normally referred to as the "general Petri net," while if "inhibitor arcs" are added it is called an "extended Petri net" (also recall that "finite capacity nets" can be reduced to general Petri nets and that marked graphs and state machines [61] are special cases of general Petri nets). If the initial marking is pre-specified, then we will refer to the Petri net as $(\mathcal{N}, M_0)$ or simply PN, whereas, if the initial marking is not specified we will refer to the net as $\mathcal{N}$. Also note that if $W(p,t) = \alpha$ (or $W(t,p) = \alpha'$) then this is often represented graphically by α (α') arcs from p to t (t to p) each with no numeric label.

Let $M_k(p_i)$ denote the marking (i.e., the number of tokens) at place $p_i \in P$ at time k and let the marking (state) of PN at time k (the "k" will be dropped when it is not needed) be denoted by

$$M_k = [M_k(p_1), \dots, M_k(p_m)]^\top.$$

A transition $t_j \in T$ is said to be *enabled* at time k if

$$M_k(p_i) \geq W(p_i, t_j)$$

for all $p_i \in P$ such that $(p_i, t_j) \in F$. It is assumed that at each time k there exists at least one transition to fire. If a transition is enabled, then it can fire. If an enabled transition $t_j \in T$ fires at time k, then the next marking for place $p_i \in P$ is given by

$$M_{k+1}(p_i) = M_k(p_i) + W(t_j, p_i) - W(p_i, t_j),$$

where $(t_j, p_i) \in F$ and $(p_i, t_j) \in F$.

Let $A = [a_{ij}]$ denote an $n \times m$ matrix of integers (the "incidence matrix"), where

$$a_{ij} = a_{ij}^+ - a_{ij}^-$$

with

$$a_{ij}^+ = W(t_i, p_j)$$

and

$$a_{ij}^- = W(p_j, t_i).$$

Let $u_k \in \{0,1\}^n$ (an n vector of zeros and ones) denote a firing vector where, if $t_j \in T$ is fired, then its corresponding firing vector is $u_k = [0, \cdots 0, 1, 0, \cdots, 0]^\top$ with the "1" in the jth position in the vector and zeros are everywhere else. The "matrix equations" (nonlinear difference equations defined on $\mathbb{N}^m$ with nonunique solutions) describing the dynamical behavior represented by a Petri net are given by

$$M_{k+1} = M_k + A^\top u_k, \tag{2.9}$$

where, if at step k,

$$a_{ij}^- \leq M_k(p_j)$$

for all $p_j \in P$, then $t_i \in T$ is enabled and if this $t_i \in T$ fires then its corresponding firing vector u_k is utilized in Equation (2.9) to generate the next state. Let $R(M_0)$ denote the set of makings of PN (states) that can be reached from M_0. Let $R_1(M)$ denote the set of all markings that are reachable from M in one transition firing. Notice that if $M_d \in R(M_0)$, and we fire some sequence of d transitions with corresponding firing vectors $u_0, u_1, u_2, \dots, u_{d-1}$ we will get

$$M_d = M_0 + A^\top u$$

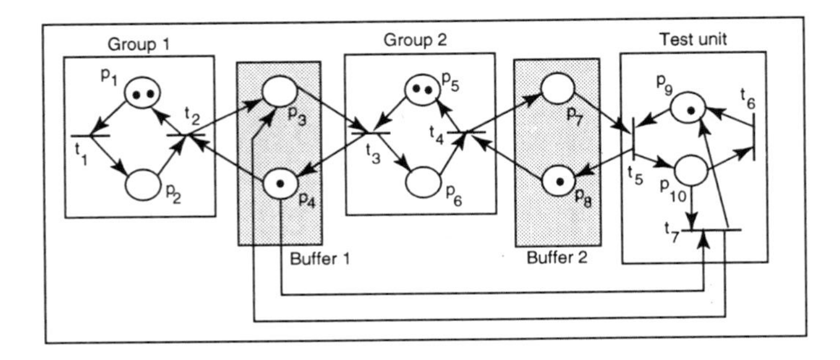

Fig. 2.4 Petri net model of a production network.

with

$$u = \sum_{k=0}^{d-1} u_k,$$

where u is called the "firing count vector."

An extended Petri net is obtained from a general Petri net by adding inhibitor arcs (sometimes called "not arcs"). Let $F_n \subset (P \times T)$ denote the set of inhibitor arcs for the extended Petri net $EPN = (P, T, F, F_n, W, M_0)$ $(F \bigcap F_n = \emptyset)$. We use a line with a small circle on the end to graphically represent the inhibitor arc. The inhibitor arc does not change in any way what happens when a transition $t \in T$ fires (i.e., Equation (2.9) holds for the extended Petri net). The inhibitor arc does, however, change which transitions are enabled at each step. The set of transitions in EPN enabled at time k is given by $\{t_j : \ M_k(p_i) \geq W(p_i, t_j) \ \text{ for all } \ p_i \in P \ \text{ s.t. } \ (p_i, t_j) \in F\} + \{t_j : (p_i, t_j) \in F_n \ \text{ and } \ M_k(p_i) = 0\}$. Hence, the inhibitor arc tests if a place has a zero marking. It is important to study properties of extended Petri nets due to fact that the addition of the inhibitor arc greatly enhances the "modeling power" of the Petri net.

2.5 PETRI NET EXAMPLES

In this section we will provide two examples of how to model systems using Petri nets. In the first we will use a general Petri net to represent a type of manufacturing system and in the second we will represent a simple computer network with an extended Petri net.

2.5.1 Production Network

Consider a production network that consists of two groups of machines, two buffers, and an inspection test unit as shown in Figure 2.4.

The marking

- $M(p_1)$ ($M(p_5)$) represents the number of parts waiting to be processed by a machine in Group 1 (2);
- $M(p_2)$ ($M(p_6)$) represents the number of parts being processed by some machine in Group 1 (2);
- $M(p_3)$ ($M(p_7)$) is the number of parts in Buffer 1 (2);
- $M(p_4)$ ($M(p_8)$) is a part counter for Buffer 1 (2); and
- $M(p_{10})$ and $M(p_9)$ are used to store parts in the inspection unit and to limit the number of parts being tested, respectively.

Transition t_1 (t_3) represents the event that some machine in Group 1 (2) begins working on a part and t_2 (t_4) represents that a machine in Group 1 (2) finishes working on a part, passes it on for further processing, and allows one more part to come in. Transition t_5 represents the event that the test unit begins inspection of a part and t_6 and t_7 represent the events that the part is accepted or rejected, respectively. As indicated, if a part is rejected, it must be returned to Buffer 1 so that Group 2 machines can perform further processing.

To study how the production network behaves the reader should choose some nonzero initial marking M_0 and study how the marking M_k evolves as various transitions fire. For example, if

$$M_0 = [2, 0, 0, 1, 2, 0, 0, 1, 1, 0]^\top$$

is the initial marking (state), as shown in Figure 2.4 and t_1 fires, then

$$M_1 = [1, 1, 0, 1, 2, 0, 0, 1, 1, 0]^\top.$$

If we have t_2 fire next, then

$$M_2 = [2, 0, 1, 0, 2, 0, 0, 1, 1, 0]^\top.$$

The sequence of t_1 firing then t_2 firing represents that a machine in Group 1 starts processing a part (t_1) and finishes it and passes it to a buffer before it enters a machine in Group 2 (t_2). Notice that the way Buffer 1 is designed p_4 serves as a buffer counter so that no more than a fixed number of parts ($M_0(p_4)$) can enter Buffer 1. The reader should consider what behavior is modeled by firing a few more transitions. Moreover, for practice, the reader should write down the incidence matrix and the matrix equation defining the nonlinear difference equation that represents the production network.

We see that the Petri net provides for a very intuitive modeling approach. The graph used for the Petri net should not be confused with the graph that is used to represent the DES model in Section 2.2. While above we show how to use the general Petri net for modeling a manufacturing problem, in the next section we will show how to use the extended Petri net to represent a computer network.

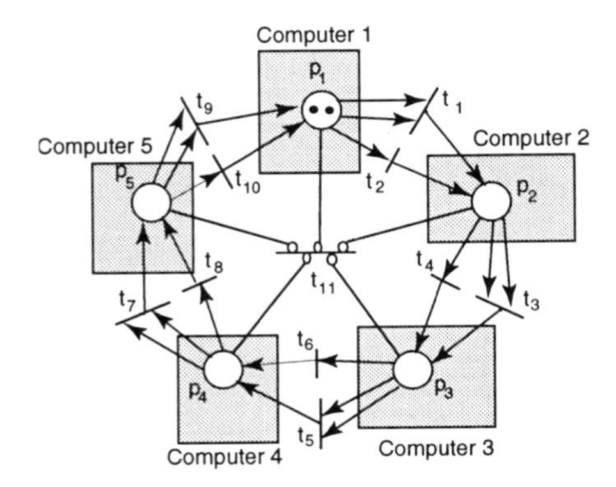

Fig. 2.5 Petri net model of a network of computers.

2.5.2 Network of Computers

Suppose that we are given a network of computers arranged in a "ring." The type of network that we consider has been considered to be useful in mutual exclusion problems, for ensuring fairness, and so on. The extended Petri net model EPN for the network of computers is shown in Figure 2.5.

Each place $p_i \in P$ represents a computer node in the network and the state of the computer (representing the possession of resources for carrying on a communication) is modeled via $M(p_i)$. The communications between the nodes are represented with the transitions (e.g., transitions t_1 and t_2 each represent different ways that node 1 can communicate to node 2). The transition t_{11} simply models a null event, where if there are no tokens in any place, then it will fire. Note that although we show only five nodes in the network, the analysis here is also valid for a general ring with N nodes.

If we start with the marking given in Figure 2.5, then fire t_2, t_4, t_6, t_8, and t_{10} we end up at the same marking as we had initially. If, however, we fire t_1 first then we can only fire t_4, t_6, t_8, t_{10}, and t_2 and then repeat this sequence over and over (passing the token around the ring for all time). Notice that for a finite nonzero initial marking if the odd numbered transitions fire enough times, eventually the network must settle into a pattern where the even numbered transitions fire in sequence an infinite number of times (the odd numbered transitions drain the network of tokens). For practice, the reader should write down the incidence matrix and the matrix equation defining the nonlinear difference equation that represents the network of computers.

2.6 SUMMARY

In this chapter we have provided an introduction to two types of DES models and have provided illustrative examples of how to represent some manufacturing systems and computer networks with these. Upon completion of this chapter the reader should understand the following topics:

- The DES model G and each of its components (the set of states, events, the enable function g and state transition function f_e, and the valid event trajectories $\boldsymbol{E}_v$).
- The concept of the set of valid and allowable event trajectories and the difference between the two.
- The concept of a state trajectory and an event trajectory.
- How to model a simple single buffer machine and a simple computer network load balancing problem.
- General and extended Petri nets (including places, transitions, markings, arcs, inhibitor arcs, etc.).
- How to define the incidence matrix and matrix equations for a Petri net.
- How to model a simple production network and a computer network with a Petri net.

Essentially, this provides a checklist for the major topics of this chapter. The reader should be sure to understand each of the above concepts or modeling approaches before moving on to the more advanced chapters.

2.7 FOR FURTHER STUDY

The DES model in Section 2.2 was introduced in [77, 81, 78]. There are many good books and articles on Petri nets, including [61, 88]. The examples for the Petri nets were taken from [80]. The production network is similar to the one in [47, 46]. The type of computer network that we consider is similar to the ones in [24], which have been considered to be useful in mutual exclusion problems, for ensuring fairness, and so on (for a more detailed explanation of their utility see [92]).

Other popular DES models include the automata model of Ramadge and Wonham [91, 90] and the models used for perturbation analysis [19]. In addition to these there are many other automata-type models, path-based models, Petri nets (e.g., colored Petri nets), and other formalisms [109] that can be used for discrete event systems. Simulation issues for discrete event systems are discussed in [108].

2.8 PROBLEMS

Problem 2.1 (DES Model for a Two-Buffer Machine): Suppose that we have a machine with two buffers, each that holds a different part type and that the machine can only process one part of one type at a time. It takes some fixed, unknown, but bounded amount of time to process a part and when processing is completed the parts are output from the system. Parts may arrive to be stored in the buffers at any finite rate. We will consider a part to be still in the buffer, even though the machine starts processing it. In this problem you will produce a discrete event system model G, which is introduced in Section 2.2 on page 9 for this machine.

(a) Define the set of states and events. Be sure to define events that represent the simultaneous arrival or departure and departure of parts at the two buffers.

(b) Define the enable and transition functions.

(c) Define the set of valid event trajectories to include all those event trajectories that result in the buffer levels being bounded by some fixed bound B.

Problem 2.2 (Petri Net Matrix Equations Definition): Write down the matrix equations (i.e., Equation (2.9)) for the production network given in Section 2.5.1 and the computer network given in Section 2.5.2.

Problem 2.3 (DES Model for a Production Network): Produce a DES model G for the production network given in Section 2.5.1. Be sure that your definition results in a nonlinear difference equation that is equivalent to the matrix equation for the Petri net model of the production network given in Section 2.5.1.

Problem 2.4 (DES Model for a Computer Network): Produce a DES model G for the computer network given in Section 2.5.2. Be sure that your definition results in a nonlinear difference equation that is equivalent to the matrix equation for the Petri net model of the computer network given in Section 2.5.2.

Problem 2.5 (Simulation of a Load Balancing System): Simulate the load balancing system in Section 2.3.2 on page 12 for the following two sets of initial conditions: $\boldsymbol{x}_0 = [0, 10]^\top$ and $\boldsymbol{x}_0 = [5, 1]^\top$. Provide plots of x^1 and x^2 vs. k for both cases.

Problem 2.6 (Petri Net Model of a Machine): In this problem you will develop Petri net models for two simple systems.

(a) Produce a Petri net model of the single buffer machine that is modeled with the DES model G in Section 2.3.1.

(b) Produce a Petri net model of the two-buffer machine that is modeled with the DES model G in Problem 2.1.

(c) Write down the incidence matrix and matrix equations (i.e., Equation (2.9)) for the models you produce in (a) and (b).

(d) Simulate the machines in (a) and (b) for various scenarios to illustrate their operation.

Problem 2.7 (Petri Net Model of a Traffic System)*:

(a) Produce a Petri net model of a typical traffic intersection that has traffic lights for two roads with traffic flowing in the north to south, south to north, east to west, and west to east directions. It should have turn lanes and you should make sure that no collisions occur; hence, you will be inherently also modeling the traffic light controller.

(b) Simulate the traffic system for various scenarios to illustrate its operation.

3

Stability Concepts and Analysis Techniques

3.1 OVERVIEW

Now that we have introduced several ways to model discrete event systems we will introduce several characterizations of stability and provide approaches to analyze stability properties. There are many types of stability. For instance, for the load balancing system in Section 2.3.2, as time evolves, the load distribution will go from being uneven to an even distribution (i.e., the buffers will eventually become nearly equal, but not necessarily perfectly equal, since that depends on the number of initial jobs in the system). This is called "asymptotic stability." In the single-buffer machine of Section 2.3.1 we use stability concepts to characterize the buffer level. For instance, if the buffer level stays bounded for any initial buffer level, then we say that the buffer level is "uniformly bounded" or that it is stable in the sense of Lagrange. There are many variations on these stability concepts and we will introduce these in this chapter.

We begin by introducing some mathematical preliminaries. Next, we provide definitions of the following stability concepts:

- Stability in the sense of Lyapunov.
- Asymptotic stability.
- Asymptotic stability in the large.
- Exponential stability.
- Exponential stability in the large.

- Stability in the sense of Lagrange.
- Uniform boundedness.
- Uniform ultimate boundedness.

Following this we will provide sufficient conditions for these properties to hold. All these conditions depend on the specification of a "Lyapunov function" that satisfies certain properties. Since all our definitions are based on a metric space formulation, we will always assume that it is possible to define a metric on the set of states for the model we consider (e.g., the model G).

We cover a few examples of stability analysis in this chapter, but refer the reader to Chapters 4–6 for the analysis of stability for more realistic applications. In particular, in this chapter we explain the relevance of stability properties and analysis to automata and finite state machine DES models. Moreover, we show how to analyze stability properties of systems modeled with Petri nets.

This chapter is central to understanding the entire book, so the reader should study it carefully. It is possible to skip the automata, finite-state machine, and Petri net material if these topics do not interest you. Moreover, you could just focus on stability in the sense of Lyapunov, asymptotic stability (in the large), and exponential stability (in the large) if you are only interested in the load balancing systems of Chapter 4. Similarly, you could focus only on the boundedness properties if you are only interested in the analysis of manufacturing systems in Chapter 5. Chapter 6 will, however, require at least a cursory understanding of most of this chapter.

3.2 MATHEMATICAL PRELIMINARIES

Let $\Re^+$ denote the nonnegative reals. Let $\rho : \mathcal{X} \times \mathcal{X} \to \Re^+$ denote a *metric* on $\mathcal{X}$, and $\{\mathcal{X}; \rho\}$ a metric space. $\{\mathcal{X}; \rho\}$ is a metric space if

- $\rho(\boldsymbol{x}, \boldsymbol{y}) = \rho(\boldsymbol{y}, \boldsymbol{x})$ for all $\boldsymbol{x}, \boldsymbol{y} \in \mathcal{X}$ and
- $\rho(\boldsymbol{x}, \boldsymbol{z}) \leq \rho(\boldsymbol{x}, \boldsymbol{y}) + \rho(\boldsymbol{y}, \boldsymbol{z})$ for all $\boldsymbol{x}, \boldsymbol{y}, \boldsymbol{z} \in \mathcal{X}$ (i.e., the triangle inequality is satisfied).

If $\boldsymbol{x} = [x_1, \ldots, x_n]^\top \in \Re^n$ and $\boldsymbol{y} = [y_1, \ldots, y_n]^\top \in \Re^n$, then some examples of valid metrics are:

- $\rho_p(\boldsymbol{x}, \boldsymbol{y}) = \left(\sum_{i=1}^n |x_i - y_i|^p\right)^{1/p}$
- $\rho_\infty(\boldsymbol{x}, \boldsymbol{y}) = \sup_{i=1,2,\ldots,n}\{|x_i - y_i|\}$.

Basically, a metric is a generalized distance measure.

Let $\mathcal{X}_z \subset \mathcal{X}$ and

$$\rho(\boldsymbol{x}, \mathcal{X}_z) = \inf\{\rho(\boldsymbol{x}, \boldsymbol{x}') : \boldsymbol{x}' \in \mathcal{X}_z\}$$

denote the distance from point $\boldsymbol{x}$ to the set $\mathcal{X}_z$. The *r-neighborhood* of an arbitrary set $\mathcal{X}_z \subset \mathcal{X}$ is denoted by the set

$$S(\mathcal{X}_z; r) = \{\boldsymbol{x} : \ 0 < \rho(\boldsymbol{x}, \mathcal{X}_z) < r\},$$

where $r > 0$. Also, let

$$\bar{S}(\mathcal{X}_z; R) = \{\boldsymbol{x} \in \mathcal{X} : \ \rho(\boldsymbol{x}, \mathcal{X}_z) \geq R\}.$$

A continuous function

$$\psi : \ [0, r_1] \rightarrow \Re^+$$

(respectively, $\psi : \ [0, \infty) \rightarrow \Re^+$) is said to belong to class K, that is, $\psi \in K$, if $\psi(0) = 0$ and if ψ is strictly increasing on $[0, r_1]$ (resp., on $[0, \infty)$). If

$$\psi : \ \Re^+ \rightarrow \Re^+,$$

if $\psi \in K$, and if

$$\lim_{r \rightarrow \infty} \psi(r) = \infty,$$

then ψ is said to belong to class KR.

A set is called *invariant* with respect to G if all motions originating in the set remain in the set. Mathematically, the set $\mathcal{X}_m \subset \mathcal{X}$ is an invariant set with respect to G if $\boldsymbol{x}_0 \in \mathcal{X}_m$ implies that $X(\boldsymbol{x}_0, E_k, k) \in \mathcal{X}_m$ for all $k \in \mathbb{N}$ and all E_k such that $E_k E \in \boldsymbol{E}_v(\boldsymbol{x}_0)$ (due to the definition of invariance, all invariant sets are closed with respect to $\{\mathcal{X}; \rho\}$). Let $\mathcal{X}_b \subset \mathcal{X}$ denote a bounded subset of $\mathcal{X}$ for the remainder of the chapter. Recall that $\boldsymbol{E}_a \subset \boldsymbol{E}_v$ is a set of allowed event trajectories.

3.3 STABILITY CONCEPTS

In this section we will provide definitions of a wide variety of stability concepts that we will study in this book.

Definition 3.1 (Stable in the Sense of Lyapunov, Asymptotic Stability): *A closed invariant set $\mathcal{X}_m \subset \mathcal{X}$ of G is called* stable in the sense of Lyapunov *with respect to $\boldsymbol{E}_a$ if for any $\epsilon > 0$ it is possible to find some $\delta > 0$ such that when $\rho(\boldsymbol{x}_0, \mathcal{X}_m) < \delta$, we have*

$$\rho(X(\boldsymbol{x}_0, E_k, k), \mathcal{X}_m) < \epsilon$$

for all E_k such that $E_k E \in \boldsymbol{E}_a(\boldsymbol{x}_0)$ and $k \geq 0$. If furthermore

$$\rho(X(\boldsymbol{x}_0, E_k, k), \mathcal{X}_m) \longrightarrow 0$$

as $k \longrightarrow \infty$, then the closed invariant set $\mathcal{X}_m$ of G is called asymptotically stable *with respect to $\boldsymbol{E}_a$.*

As is always the case, the properties in Definition 3.1 are *local* stability properties, that is, with respect to some r-neighborhood. An invariant set $\mathcal{X}_m$ is stable in the sense of Lyapunov if for every given $\epsilon > 0$ you can find a δ-neighborhood of $\mathcal{X}_m$ where if you start the motion in the δ-neighborhood, the motion will stay in the ϵ-neighborhood of $\mathcal{X}_m$. It is asymptotically stable if, furthermore, the motion converges to $\mathcal{X}_m$ no matter where it started in the δ-neighborhood. A closed invariant set $\mathcal{X}_m \subset \mathcal{X}$ of G is called *unstable* in the sense of Lyapunov w.r.t. $\boldsymbol{E}_a$ if it is not stable in the sense of Lyapunov w.r.t $\boldsymbol{E}_a$.

Definition 3.2 (Region of Asymptotic Stability, Asymptotic Stability in the Large): *If the closed invariant set $\mathcal{X}_m \subset \mathcal{X}$ of G is asymptotically stable with respect to $\boldsymbol{E}_a$, then the set $\mathcal{X}_a \subset \mathcal{X}$ having the property that for all $\boldsymbol{x}_0 \in \mathcal{X}_a$,*

$$\rho(X(\boldsymbol{x}_0, E_k, k), \mathcal{X}_m) \longrightarrow 0$$

for all E_k such that $E_k E \in \boldsymbol{E}_a(\boldsymbol{x}_0)$ as $k \longrightarrow \infty$ is called the region of asymptotic stability *of $\mathcal{X}_m$ with respect to $\boldsymbol{E}_a$. If $\mathcal{X}_a = \mathcal{X}$, then the closed invariant set $\mathcal{X}_m$ of G is called* asymptotically stable in the large *with respect to $\boldsymbol{E}_a$.*

The region of asymptotic stability is simply the region around $\mathcal{X}_m$ for which the motions converge to $\mathcal{X}_m$. In addition to our concern that eventually

$$\rho(X(\boldsymbol{x}_0, E_k, k), \mathcal{X}_m) \longrightarrow 0,$$

we may be concerned with how *quickly* any state trajectory must reach the invariant set. This is the basic reason for defining "exponential stability" as we do next.

Definition 3.3 (Exponential Stability): *We say that the closed invariant set $\mathcal{X}_m \subset \mathcal{X}$ of G is* exponentially stable *with respect to $\boldsymbol{E}_a$ if for $\boldsymbol{x}_0 \in S(\mathcal{X}_m; r)$,*

$$\rho(X(\boldsymbol{x}_0, E_k, k), \mathcal{X}_m) \leq \zeta e^{-\alpha k} \rho(\boldsymbol{x}_0, \mathcal{X}_m)$$

for some $\alpha > 0$ and some $\zeta > 0$ and for all E_k such that $E_k E \in \boldsymbol{E}_a(\boldsymbol{x}_0)$ and $k \geq 0$.

Exponential stability is similar to asymptotic stability, except it is a stronger form of stability since it helps to characterize the rate at which motions converge to $\mathcal{X}_m$. Just like for asymptotic stability, we have a region of exponential stability.

Definition 3.4 (Region of Exponential Stability, Exponential Stability in the Large): *If the closed invariant set $\mathcal{X}_m \subset \mathcal{X}$ of G is* *exponentially stable with respect to* $\boldsymbol{E}_a$*, then the set* $\mathcal{X}_e \subset \mathcal{X}$ *having the property that for all* $\boldsymbol{x}_0 \in \mathcal{X}_e$*,*

$$\rho(X(\boldsymbol{x}_0, E_k, k), \mathcal{X}_m) \leq \zeta e^{-\alpha k} \rho(\boldsymbol{x}_0, \mathcal{X}_m)$$

for some $\alpha > 0$ *and some* $\zeta > 0$*, for all* E_k *and* $k \geq 0$ *such that* $E_k E \in \boldsymbol{E}_a(\boldsymbol{x}_0)$ *is called the* region of exponential stability *of* $\mathcal{X}_m$ *with respect to* $\boldsymbol{E}_a$*. If* $\mathcal{X}_e = \mathcal{X}$*, then the closed invariant set* $\mathcal{X}_m$ *of* G *is called* exponentially stable in the large *with respect to* $\boldsymbol{E}_a$*.*

Note that the above characterizations of stability are quite general. For instance, let $\mathcal{X}_0$ denote a set of possible initial states and let $\mathcal{X}_m$ contain the elements of all the motions $X(x_0, E_k, k)$ such that $x_0 \in \mathcal{X}_0$ and E_k satisfies $E_k E \in \boldsymbol{E}_a(x_0)$, where E is an infinite event sequence. Studying the stability of the invariant set $\mathcal{X}_m$ is similar to the study of "orbital stability" in [35]. For this invariant set $\mathcal{X}_m$ it could also be assumed that each of these motions visits some pre-specified set $\mathcal{X}_s \subset \mathcal{X}_m$ infinitely often or that the motions satisfy some other property. This shows one connection between the work in temporal logic and automata-theoretic studies and Lyapunov stability analysis. More such connections will be highlighted in Section 3.6.1.

Next, we will discuss some boundedness properties and alternative stability definitions.

Definition 3.5 (Boundedness, Lagrange Stability): *The motions of* G *which begin at* $\boldsymbol{x}_0 \in \mathcal{X}$ *(*$X(\boldsymbol{x}_0, E_k, k)$*) are* bounded w.r.t $\boldsymbol{E}_a$ *and* $\mathcal{X}_b$ *if there exists a* $\beta > 0$ *such that*

$$\rho(X(\boldsymbol{x}_0, E_k, k), \mathcal{X}_b) < \beta$$

for all E_k *such that* $E_k E \in \boldsymbol{E}_a(\boldsymbol{x}_0)$ *and for all* $k \geq 0$*. The DES* G *is said to possess* Lagrange Stability *w.r.t.* $\boldsymbol{E}_a$ *and* $\mathcal{X}_b$ *if for each* $\boldsymbol{x}_0 \in \mathcal{X}$ *the motions* $X(\boldsymbol{x}_0, E_k, k)$ *for all* E_k *such that* $E_k E \in \boldsymbol{E}_a(\boldsymbol{x}_0)$ *and all* $k \geq 0$ *are bounded w.r.t.* $\boldsymbol{E}_a$ *and* $\mathcal{X}_b$*.*

Boundedness and Lagrange stability simply characterize the boundedness of the motions. Two more ways to characterize boundedness are given next.

Definition 3.6 (Uniform Boundedness): *The motions of* G *are* uniformly bounded w.r.t $\boldsymbol{E}_a$ *and* $\mathcal{X}_b$ *if for any* $\alpha > 0$ *there exists a* $\beta > 0$ *(that depends on* α*), such that if* $\rho(\boldsymbol{x}_0, \mathcal{X}_b) < \alpha$ *then*

$$\rho(X(\boldsymbol{x}_0, E_k, k), \mathcal{X}_b) < \beta$$

for all E_k*, such that* $E_k E \in \boldsymbol{E}_a(\boldsymbol{x}_0)$ *and for all* $k \geq 0$*.*

Definition 3.7 (Uniform Ultimate Boundedness): *The motions of* G *are* uniformly ultimately bounded with bound B w.r.t $\boldsymbol{E}_a$ *and* $\mathcal{X}_b$ *if* *there exists a* $B > 0$ *and if corresponding to any* $\alpha > 0$ *there exists* $T(\alpha) > 0$*, such that* $\rho(\boldsymbol{x}_0, \mathcal{X}_b) < \alpha$ *implies that*

$$\rho(X(\boldsymbol{x}_0, E_{k'}, k'), \mathcal{X}_b) < B$$

for all $E_{k'}$*, such that* $E_{k'}E \in \boldsymbol{E}_a(\boldsymbol{x}_0)$ *where* $k' \geq T(\alpha)$*.*

Uniform ultimate boundedness simply says if you start a motion in an α-neighborhood, it will eventually (ultimately) enter a B-neighborhood (so in an initial transient period the size of the motion can be larger than B).

While the above definitions provide characterizations of the standard stability properties, there are many other ways to think about stability. For instance, next we provide two additional ways to characterize stability properties.

Definition 3.8 (Practical Stability): *Fix* α *and* β *such that* $\beta \geq \alpha > 0$*, let* ρ *be a specified metric on* $\mathcal{X}$*, and let* $\mathcal{X}_b \subset \mathcal{X}$ *and* $\boldsymbol{E}_a \subset \boldsymbol{E}_v$*. The DES* G *is said to be* practically stable w.r.t. $(\alpha, \beta, \rho, \mathcal{X}_b, \boldsymbol{E}_a)$ *if for all* $\boldsymbol{x}_0 \in \mathcal{X}$ *such that* $\rho(\boldsymbol{x}_0, \mathcal{X}_b) < \alpha$*,* $\rho(X(\boldsymbol{x}_0, E_k, k), \mathcal{X}_b) < \beta$ *for all* E_k *such that* $E_k E \in \boldsymbol{E}_a(\boldsymbol{x}_0)$ *and all* $k \geq 0$*.*

Definition 3.9 (Finite Time Stability): *Fix* α *and* β *such that* $\beta \geq \alpha > 0$*, let* ρ *be a specified metric on* $\mathcal{X}$*, and let* $\mathcal{X}_b \subset \mathcal{X}$ *and* $\boldsymbol{E}_a \subset \boldsymbol{E}_v$*. Furthermore, let* T_f *denote a fixed* final time. *The DES* G *is said to be* finite-time stable w.r.t. $(\alpha, \beta, T_f, \rho, \mathcal{X}_b, \boldsymbol{E}_a)$ *if for all* $\boldsymbol{x}_0 \in \mathcal{X}$ *such that* $\rho(\boldsymbol{x}_0, \mathcal{X}_b) < \alpha$*,* $\rho(X(\boldsymbol{x}_0, E_{k'}, k'), \mathcal{X}_b) < \beta$ *for all* $E_{k'}$ *such that* $E_{k'}E \in \boldsymbol{E}_a(\boldsymbol{x}_0)$ *where* $k' < T_f$*.*

Notice that if the above stability and boundedness properties hold for some $\boldsymbol{E}_a$ then they also hold for all $\boldsymbol{E}'_a$ such that $\boldsymbol{E}'_a \subset \boldsymbol{E}_a$.

3.4 STABILITY ANALYSIS TECHNIQUES

In this section we provide sufficient conditions for the stability properties of the previous section to hold. It is these sufficient conditions that will be used in the remainder of the book to analyze stability properties of DES. Basically, we use a Lyapunov framework and the function V below is traditionally called the "Lyapunov function."

Theorem 3.1 (Stability in the Sense of Lyapunov): *In order for the invariant set* $\mathcal{X}_m$ *to be stable in the sense of Lyapunov w.r.t.* $\boldsymbol{E}_a$ *it is sufficient that in a neighborhood* $S(\mathcal{X}_m; r)$ *there exists a specified functional* V *and* $\psi_1, \psi_2 \in K$ *such that:*

(*i*) $\psi_1(\rho(\boldsymbol{x}, \mathcal{X}_m)) \leq V(\boldsymbol{x}) \leq \psi_2(\rho(\boldsymbol{x}, \mathcal{X}_m))$*, and*

(ii) *$V(X(\boldsymbol{x}_0, E_k, k))$ is a nonincreasing function for $\boldsymbol{x}_0 \in S(\mathcal{X}_m; r)$, for all E_k such that $E_k E \in \boldsymbol{E}_a(\boldsymbol{x}_0)$ and all $k \geq 0$ (i.e., V is nonincreasing along all possible motions of the system).*

Proof: Fix $\epsilon > 0$, such that $\epsilon < r$. Pick $\delta > 0$ so small that $\psi_2(\delta) < \psi_1(\epsilon)$. If $\rho(\boldsymbol{x}_0, \mathcal{X}_m) \leq \delta$ then $\psi_2(\rho(\boldsymbol{x}_0, \mathcal{X}_m)) \leq \psi_2(\delta)$ so

$$V(\boldsymbol{x}_0) \leq \psi_2(\delta) < \psi_1(\epsilon).$$

Since $V(X(\boldsymbol{x}_0, E_k, k))$ is a nonincreasing function $V(X(\boldsymbol{x}_0, E_k, k)) < \psi_1(\epsilon)$ for all E_k such that $E_k E \in \boldsymbol{E}_a(\boldsymbol{x}_0)$ and all $k \geq 0$. Hence, since

$$\psi_1(\rho(X(\boldsymbol{x}_0, E_k, k), \mathcal{X}_m)) \leq V(X(\boldsymbol{x}_0, E_k, k))$$

we know $\psi_1(\rho(X(\boldsymbol{x}_0, E_k, k), \mathcal{X}_m)) < \psi_1(\epsilon)$ for all E_k such that $E_k E \in \boldsymbol{E}_a(\boldsymbol{x}_0)$ and all $k \geq 0$. Since $\psi_1 \in K$ is monotone increasing its inverse exists so we know that $\rho(X(\boldsymbol{x}_0, E_k, k), \mathcal{X}_m) < \epsilon$ for all E_k, such that $E_k E \in \boldsymbol{E}_a(\boldsymbol{x}_0)$ and all $k \geq 0$. ■

Theorem 3.2 (Asymptotic Stability): *In order for the invariant set $\mathcal{X}_m$ to be asymptotically stable w.r.t. $\boldsymbol{E}_a$ it is sufficient that in a neighborhood $S(\mathcal{X}_m; r)$ there exists a specified functional V and $\psi_1, \psi_2 \in K$ that satisfy the properties of Theorem 3.1 and there exists $\psi_3 \in K$ such that*

$$V(X(\boldsymbol{x}_0, E_{k+1}, k+1)) - V(X(\boldsymbol{x}_0, E_k, k)) \leq -\psi_3(\rho(X(\boldsymbol{x}_0, E_k, k), \mathcal{X}_m))$$

for all $\boldsymbol{x}_0 \in S(\mathcal{X}_m; r)$ and for all E_k such that $E_{k+1} = E_k e$ ($e \in \mathcal{E}$) and $E_{k+1}E \in \boldsymbol{E}_a(\boldsymbol{x}_0)$ and all $k \geq 0$.

Proof: By Theorem 3.1 we know that $\mathcal{X}_m$ is stable in the sense of Lyapunov w.r.t. $\boldsymbol{E}_a$. Pick $\delta_1 > 0$ such that $\psi_2(\delta_1) < \psi_1(r)$. Choose ϵ such that $0 < \epsilon < r$. Choose δ_2 such that $0 < \delta_2 < \delta_1$ and such that $\psi_2(\delta_2) < \psi_1(\epsilon)$. Pick $T = \psi_1(r)/\psi_3(\delta_2)$. Choose $\boldsymbol{x}_0$ such that $\rho(\boldsymbol{x}_0, \mathcal{X}_m) < \delta_1$. It must be that $\rho(X(\boldsymbol{x}_0, E_k, k), \mathcal{X}_m) < \delta_2$ for some $k^* \in [0, T]$ for if this were not true we would have $\rho(X(\boldsymbol{x}_0, E_k, k), \mathcal{X}_m) \geq \delta_2$ for all $k \in [0, T]$. If this were the case then

$$0 < \psi_1(\delta_2) \leq V(X(\boldsymbol{x}_0, E_k, k))$$

But,

$$V(X(\boldsymbol{x}_0, E_k, k)) = V(\boldsymbol{x}_0) + \sum_{j=0}^{k-1} \left(V(X(\boldsymbol{x}_0, E_{j+1}, j+1)) - V(X(\boldsymbol{x}_0, E_j, j))\right).$$

Now, notice that

$$V(\boldsymbol{x}_0) + \sum_{j=0}^{k-1} \left(V(X(\boldsymbol{x}_0, E_{j+1}, j+1)) - V(X(\boldsymbol{x}_0, E_j, j))\right) \leq \psi_2(\delta_1) - \sum_{j=0}^{k-1} \psi_3(\delta_2).$$

At time $k = T$ we would also have

$$0 < \psi_2(\delta_1) - T\psi_3(\delta_2) = \psi_2(\delta_1) - \psi_1(r)$$

which is a contradiction. Hence k^* exists. Now, for $k \geq k^*$ we have

$$\psi_1(\rho(X(\boldsymbol{x}_0, E_k, k), \mathcal{X}_m)) \leq V(X(\boldsymbol{x}_0, E_k, k)) \leq V(X(\boldsymbol{x}_0, E_{k^*}, k^*))$$

and

$$V(X(\boldsymbol{x}_0, E_{k^*}, k^*)) \leq \psi_2(\rho(X(\boldsymbol{x}_0, E_{k^*}, k^*), \mathcal{X}_m)) \leq \psi_2(\delta_2) < \psi_1(\epsilon).$$

Since $\psi_1 \in K$, ψ_1^{-1} exists so using the two previous inequalities

$$\rho(X(\boldsymbol{x}_0, E_k, k), \mathcal{X}_m) < \epsilon$$

for all $k \geq k^*$ and hence for all $k \geq T$ and this guarantees that

$$\rho(X(\boldsymbol{x}_0, E_k, k), \mathcal{X}_m) \longrightarrow 0$$

as $k \longrightarrow \infty$ for all E_k, such that $E_k E \in \boldsymbol{E}_a(\boldsymbol{x}_0)$ and all $k \geq 0$. ■

If the properties in the statement of Theorem 3.2 hold over all of $\mathcal{X}$ and $\psi_1, \psi_2, \psi_3 \in KR$, then the invariant set $\mathcal{X}_m$ is asymptotically stable in the large w.r.t. $\boldsymbol{E}_a$.

Theorem 3.3 (Exponential Stability): *In order for the invariant set $\mathcal{X}_m$ to be exponentially stable w.r.t. $\boldsymbol{E}_a$ it is sufficient that in a neighborhood $S(\mathcal{X}_m; r)$ there exists a specified functional V and three positive constants c_1, c_2, and c_3 such that $c_2 > c_3$ and*

(i) $c_1\rho(\boldsymbol{x}, \mathcal{X}_m) \leq V(\boldsymbol{x}) \leq c_2\rho(\boldsymbol{x}, \mathcal{X}_m)$, *and*

(ii) $V(X(\boldsymbol{x}_0, E_{k+1}, k+1)) - V(X(\boldsymbol{x}_0, E_k, k)) \leq -c_3\rho(X(\boldsymbol{x}_0, E_k, k), \mathcal{X}_m)$ *for all* $\boldsymbol{x}_0 \in S(\mathcal{X}_m; r)$ *and for all* E_k *such that* $E_{k+1} = E_k e$ $(e \in \mathcal{E})$ *and* $E_{k+1}E \in \boldsymbol{E}_a(\boldsymbol{x}_0)$ *and all* $k \geq 0$.

Proof: Given $\boldsymbol{x}_0 \in S(\mathcal{X}_m; r)$ let $X_0(k) = X(\boldsymbol{x}_0, E_k, k)$ for any E_k such that $E_k E \in \boldsymbol{E}_a(\boldsymbol{x}_0)$ and all $k \geq 0$. Let $\bar{V}(k) = V(X_0(k))$. $\bar{V}(k)$ satisfies $\bar{V}(k+1) - \bar{V}(k) \leq -c_3\rho(X_0(k), \mathcal{X}_m)$. Now, since $(1/c_2)\bar{V}(k) \leq \rho(X_0(k), \mathcal{X}_m)$ for all k, we know that $\bar{V}(k+1) - \bar{V}(k) \leq -(c_3/c_2)\bar{V}(k)$. Hence,

$$\bar{V}(k+1) \leq \left(1 - \frac{c_3}{c_2}\right)\bar{V}(k)$$

or

$$\bar{V}(k) \leq \left(1 - \frac{c_3}{c_2}\right)^k \bar{V}(0)$$

so

$$c_1\rho(X_0(k),\mathcal{X}_m) \le \left(1-\frac{c_3}{c_2}\right)^k \bar{V}(0).$$

Now,

$$\bar{V}(0) \le c_2\rho(X_0(0),\mathcal{X}_m)$$

so

$$c_1\rho(X_0(k),\mathcal{X}_m) \le c_2\left(1-\frac{c_3}{c_2}\right)^k \rho(X_0(0),\mathcal{X}_m)$$

and

$$\rho(X_0(k),\mathcal{X}_m) \le \frac{c_2}{c_1}\left(1-\frac{c_3}{c_2}\right)^k \rho(X_0(0),\mathcal{X}_m)$$

so that there exists an α and ζ so that

$$\rho(X_0(k),\mathcal{X}_m) \le \zeta e^{-\alpha k}\rho(X_0(0),\mathcal{X}_m)$$

and hence the system is exponentially stable. ∎

If the properties of Theorem 3.3 hold on all of $\mathcal{X}$, then the invariant set $\mathcal{X}_m$ is exponentially stable in the large w.r.t. $\boldsymbol{E}_a$.

Theorem 3.4 (Uniform Boundedness): *In order for the motions of G to be uniformly bounded w.r.t. $\boldsymbol{E}_a$ and $\mathcal{X}_b$ (a bounded set) it is sufficient that there exists a function V defined on $\bar{S}(\mathcal{X}_b;R)$ (where R may be large), and $\psi_1,\psi_2 \in KR$ such that*

(*i*) $\psi_1(\rho(\boldsymbol{x},\mathcal{X}_b)) \le V(\boldsymbol{x}) \le \psi_2(\rho(\boldsymbol{x},\mathcal{X}_b))$, $\boldsymbol{x}\in\bar{S}(\mathcal{X}_b;R)$, *and*

(*ii*) $V(X(\boldsymbol{x}_0,E_k,k))$ *is a nonincreasing function for* $\boldsymbol{x}_0 \in \bar{S}(\mathcal{X}_b;R)$, *for all* E_k *such that* $E_kE \in \boldsymbol{E}_a(\boldsymbol{x}_0)$ *and all* $k \ge 0$ *(i.e.,* V *is nonincreasing along all possible motions of the system).*

Proof: Fix $r' > R$ and let $\boldsymbol{x}_0 \in S(\mathcal{X}_b;r')$ with $\rho(\boldsymbol{x}_0,\mathcal{X}_b) > R$. By conditions (i) and (ii),

$$V(X(\boldsymbol{x}_0,E_k,k)) \le V(X(\boldsymbol{x}_0,\emptyset,0)) \le \psi_2(r')$$

for all E_k, such that $E_kE \in \boldsymbol{E}_a(\boldsymbol{x}_0)$. By condition (i) it is the case that

$$\psi_1(\rho(X(\boldsymbol{x}_0,E_k,k),\mathcal{X}_b)) \le V(X(\boldsymbol{x}_0,E_k,k)) \le \psi_2(r')$$

for all E_k, such that $E_kE \in \boldsymbol{E}_a(\boldsymbol{x}_0)$ provided that $\rho(X(\boldsymbol{x}_0,E_k,k),\mathcal{X}_b) > R$. Since $\psi_1 \in KR$, its inverse exists, so

$$\rho(X(\boldsymbol{x}_0,E_k,k),\mathcal{X}_b)) \le \psi_1^{-1}(\psi_2(r')) \doteq \beta$$

for all E_k, such that $E_kE \in \boldsymbol{E}_a(\boldsymbol{x}_0)$ provided that $\rho(X(\boldsymbol{x}_0,E_k,k),\mathcal{X}_b) > R$. If $\boldsymbol{x}_0 \notin \bar{S}(\mathcal{X}_b;R)$ or if $\boldsymbol{x}_0 \in \bar{S}(\mathcal{X}_b;R)$ and there exists k', $E_{k'}$ such that $E_{k'}E \in$

$\boldsymbol{E}_a(\boldsymbol{x}_0)$ where $X(\boldsymbol{x}_0, E_{k'}, k') \notin \bar{S}(\mathcal{X}_b; R)$, then it could be that for all $k \geq k'$, $X(\boldsymbol{x}_0, E_k, k) \in S(\mathcal{X}_b; r')$ or it could be that for this $E_{k'}$ there exist $k_1 \geq k'$, $k_2 \geq k'$, such that $X(\boldsymbol{x}_0, E_{k''}, k'') \notin S(\mathcal{X}_b; r')$ for all k'', $k_1 < k'' < k_2 \leq \infty$. However, the above argument yields $\rho(X(\boldsymbol{x}_0, E_{k''}, k''), \mathcal{X}_b) \leq \beta$ for all such k'', so that

$$\rho(X(\boldsymbol{x}_0, E_k, k), \mathcal{X}_b)) \leq \max\{R, \beta\}$$

for all E_k, such that $E_k E \in \boldsymbol{E}_a(\boldsymbol{x}_0)$. ■

Theorem 3.5 (Uniform Ultimate Boundedness): *In order for the motions of G to be uniformly ultimately bounded with bound B w.r.t. $\boldsymbol{E}_a$ and $\mathcal{X}_b$ it is sufficient that there exists a function V defined on $\bar{S}(\mathcal{X}_b; R)$ (where R may be large), ψ_1, $\psi_2 \in KR$, and $\psi_3 \in K$ such that*

(*i*) $\psi_1(\rho(\boldsymbol{x}, \mathcal{X}_b)) \leq V(\boldsymbol{x}) \leq \psi_2(\rho(\boldsymbol{x}, \mathcal{X}_b))$, $\boldsymbol{x} \in \bar{S}(\mathcal{X}_b; R)$, *and*

(*ii*) $V(X(\boldsymbol{x}_0, E_{k+1}, k+1)) - V(X(\boldsymbol{x}_0, E_k, k)) \leq -\psi_3(\rho(X(\boldsymbol{x}_0, E_k, k), \mathcal{X}_b))$ *for all* $\boldsymbol{x}_0 \in \bar{S}(\mathcal{X}_b; R)$, *and for all* E_k *such that* $E_{k+1} = E_k e$ $(e \in \mathcal{E})$ *and* $E_{k+1}E \in \boldsymbol{E}_a(\boldsymbol{x}_0)$ *and all* $k \geq 0$.

Proof: Fix $r_1 > R$, choose $B > r_1$ such that $\psi_2(r_1) < \psi_1(B)$ (which is always possible), choose $r_2 > B$, and let

$$T' = (\psi_2(r_2)/\psi_3(r_1)) + 1.$$

With $B < \rho(\boldsymbol{x}_0, \mathcal{X}_b) \leq r_2$, assume that $\rho(X(\boldsymbol{x}_0, E_k, k), \mathcal{X}_b) > r_1$ for all E_k such that $E_k E \in \boldsymbol{E}_a(\boldsymbol{x}_0)$. By condition (*ii*),

$$V(X(\boldsymbol{x}_0, E_k, k)) \leq V(X(\boldsymbol{x}_0, \emptyset, 0)) - \sum_{j=0}^{k-1} \psi_3(\rho(X(\boldsymbol{x}_0, E_j, j), \mathcal{X}_b))$$

for all E_k, such that $E_k E \in \boldsymbol{E}_a(\boldsymbol{x}_0)$. But

$$V(X(\boldsymbol{x}_0, \emptyset, 0)) \leq \psi_2(\rho(\boldsymbol{x}_0, \mathcal{X}_b)) \leq \psi_2(r_2)$$

and $\psi_3(\rho(X(\boldsymbol{x}_0, E_j, j), \mathcal{X}_b)) > \psi_3(r_1)$, so that we get

$$0 < V(X(\boldsymbol{x}_0, E_k, k)) \leq \psi_2(r_2) - k\psi_3(r_1)$$

for all E_k, such that $E_k E \in \boldsymbol{E}_a(\boldsymbol{x}_0)$. Let

$$k = T' = (\psi_2(r_2)/\psi_3(r_1)) + 1$$

as above so that, $V(X(\boldsymbol{x}_0, E_k, k)) \leq -\psi_3(r_1)$ for all E_k, such that $E_k E \in \boldsymbol{E}_a(\boldsymbol{x}_0)$ which is a contradiction, since

$$V(X(\boldsymbol{x}_0, E_k, k)) \geq \psi_1(\rho(X(\boldsymbol{x}_0, E_k, k), \mathcal{X}_b)) \geq \psi_1(r_1).$$

Hence, there exists k^* such that

$$\rho(X(\boldsymbol{x}_0, E_{k^*}, k^*), \mathcal{X}_b) \leq r_1,$$

where $E_{k^*}E \in \boldsymbol{E}_a(\boldsymbol{x}_0)$. Suppose now that $\rho(X(\boldsymbol{x}_0, E_{k^*}, k^*), \mathcal{X}_b) \leq r_1$ and $\rho(X(\boldsymbol{x}_0, E_k, k), \mathcal{X}_b) > r_1$ for k, such that $k^* < k \leq k' \leq \infty$ and $E_kE \in \boldsymbol{E}_a(\boldsymbol{x}_0)$. Then

$$\psi_1(\rho(X(\boldsymbol{x}_0, E_k, k), \mathcal{X}_b)) \leq V(X(\boldsymbol{x}_0, E_k, k)) \leq V(X(\boldsymbol{x}_0, E_{k^*}, k^*))$$

Also,

$$V(X(\boldsymbol{x}_0, E_{k^*}, k^*)) \leq \psi_2(\rho(X(\boldsymbol{x}_0, E_{k^*}, k^*), \mathcal{X}_b))$$

and

$$\psi_2(\rho(X(\boldsymbol{x}_0, E_{k^*}, k^*), \mathcal{X}_b)) \leq \psi_2(r_1) < \psi_1(B)$$

so that, using the three above inequalities,

$$\rho(X(\boldsymbol{x}_0, E_k, k), \mathcal{X}_b) < \psi_1^{-1}(\psi_1(B)) = B$$

for all $k \geq k^*$, and E_k, such that $E_kE \in \boldsymbol{E}_a(\boldsymbol{x}_0)$. ■

Corollary 1: In order for the motions of G to be uniformly ultimately bounded with bound B w.r.t. $\boldsymbol{E}_a$ and $\mathcal{X}_b$, it is sufficient that there exists a function V defined on $\bar{S}(\mathcal{X}_b; R)$ (where R may be large), $D \in \mathbb{N}$, and $\psi_1, \psi_2 \in KR$, $\psi_3 \in K$, such that

(*i*) Conditions (*i*) and (*ii*) of Theorem 3.4 hold, and

(*ii*) $V(X(\boldsymbol{x}_0, E_{k+1}, k+1)) - V(X(\boldsymbol{x}_0, E_k, k)) \leq -\psi_3(\rho(X(\boldsymbol{x}_0, E_k, k), \mathcal{X}_b))$ for all $\boldsymbol{x}_0 \in \bar{S}(\mathcal{X}_b; R)$, and for all E_k, such that $E_{k+1} = E_k e$ ($e \in \mathcal{E}$), $E_{k+1}E \in \boldsymbol{E}_a(\boldsymbol{x}_0)$, $k \in [0, D)$, and if this inequality holds for $k' \geq 0$, then it holds for each E_k, such that $E_kE \in \boldsymbol{E}_a(\boldsymbol{x}_0)$ for some $k \in (k', k'+D]$ (i.e., for each $E \in \boldsymbol{E}_a(\boldsymbol{x}_0)$ the inequality holds at least once every D steps).

Proof: Choose r_1, r_2, and B as above and T' as above and by condition (*ii*) for $k' \geq kD$, $k \geq 0$,

$$V(X(\boldsymbol{x}_0, E_{k'}, k')) \leq V(X(\boldsymbol{x}_0, \emptyset, 0)) - k\psi_3(\rho(\boldsymbol{x}_0, \mathcal{X}_b))$$

for all $E_{k'}$, such that $E_{k'}E \in \boldsymbol{E}_a(\boldsymbol{x}_0)$. As above, we find that for $k' \geq kD$, $k \geq 0$,

$$V(X(\boldsymbol{x}_0, E_{k'}, k')) \leq \psi_2(r_2) - k\psi_3(r_1)$$

for all $E_{k'}$, such that $E_{k'}E \in \boldsymbol{E}_a(\boldsymbol{x}_0)$. Choosing $k = T'$ we get a contradiction for $k' \geq DT'$. The remainder of the proof is the same as for Theorem 3.5. ■

Using ideas from the proof for uniform boundedness (Theorem 3.4) we state and prove the following result on Lagrange stability.

Theorem 3.6 (Lagrange Stability): *For a DES G to possess Lagrange stability w.r.t. $\boldsymbol{E}_a$ and $\mathcal{X}_b$ it is sufficient that there exists a function V defined on $\mathcal{X}$ and $\psi_1, \psi_2 \in KR$ such that*

(i) $\psi_1(\rho(\boldsymbol{x}, \mathcal{X}_b)) \leq V(\boldsymbol{x}) \leq \psi_2(\rho(\boldsymbol{x}, \mathcal{X}_b))$, *for all* $\boldsymbol{x} \in \mathcal{X}$, *and*

(ii) $V(X(\boldsymbol{x}_0, E_k, k)) - V(\boldsymbol{x}_0) < \beta(\boldsymbol{x}_0)$ *for each* $\boldsymbol{x}_0 \in \mathcal{X}$, *and all* E_k, *such that* $E_k E \in \boldsymbol{E}_a(\boldsymbol{x}_0)$ *for all* $k \geq 0$ *and some* $\beta(\boldsymbol{x}_0) > 0$.

Proof: Fix $r' > 0$ and let $\boldsymbol{x}_0 \in S(\mathcal{X}_b; r')$ so that $V(\boldsymbol{x}_0) \leq \psi_2(r')$. For all E_k, such that $E_k E \in \boldsymbol{E}_a(\boldsymbol{x}_0)$ and all $k \geq 0$,

$$\psi_1(\rho(X(\boldsymbol{x}_0, E_k, k), \mathcal{X}_b)) \leq V(X(\boldsymbol{x}_0, E_k, k)) \leq V(\boldsymbol{x}_0) + \beta(\boldsymbol{x}_0) \leq \psi_2(r') + \beta(\boldsymbol{x}_0)$$

Since $\psi_1 \in KR$,

$$\rho(X(\boldsymbol{x}_0, E_k, k), \mathcal{X}_b) \leq \psi_1^{-1}(\psi_2(r') + \beta(\boldsymbol{x}_0)) \doteq \beta'(\boldsymbol{x}_0),$$

which shows that G possesses Lagrange stability. ∎

This completes our overview of stability analysis approaches for the stability concepts of the previous section. Next, we study some illustrative DES applications of the above theory.

3.5 REACHABILITY AND CYCLIC PROPERTIES

While the main focus of this book is on stability analysis, there are other properties of DES that are important to study. Two of these are reachability and cyclic properties. Next, we will define these properties more carefully. For $\mathcal{X}_m \subset \mathcal{X}$, let $\mathcal{X}(G, x_0, \mathcal{X}_m)$ denote the set of all finite length state trajectories that begin at x_0 and end in $\mathcal{X}_m$. A system G is said to be "$(x_0, \mathcal{X}_m) -$ *reachable*" if there exists a sequence of events to occur that produces a state trajectory $s \in \mathcal{X}(G, x_0, \mathcal{X}_m)$. Search algorithms can be used to find the paths to test reachability properties [73, 72]. We will analyze the reachability properties of a tank and flexible manufacturing system in Chapter 6.

Let $\mathcal{X}_c \subset \mathcal{X}$ denote a subset of the states such that each $x_c \in \mathcal{X}_c$ lies on a cycle that is in $\mathcal{X}_c$. A system G is said to be "$(x_0, \mathcal{X}_c) -$ *cyclic*" if there exists a sequence of events to occur that produces a state trajectory $s \in \mathcal{X}(G, x_0, \mathcal{X}_c)$. It is a hard problem to detect the presence of cyclic behavior in the system, since one may not be able to find $\mathcal{X}_c$ without studying all system trajectories. To help automate the testing cyclic behavior we can use a two step approach. First we specify a set $\mathcal{X}_c$ (which can sometimes be found with a search algorithm), then we use a search algorithm to find the inference path that starts at x_0^s and ends in $\mathcal{X}_c$ (if one exists) [71]. We will analyze the cyclic properties for an application in Chapter 6.

3.6 DES APPLICATIONS

In this section we show how the results of previous section can be used to characterize and analyze the stability properties of systems represented by automata-theoretic models like the "generator" in [91], general and extended Petri nets, and finite-state systems. This analysis helps to show (1) the relevance of Lyapunov stability to general "logical" DES models, and (2) some limitations of the Lyapunov stability analysis approach. Moreover, we use a rate-synchronized manufacturing line to illustrate how to analyze boundedness properties for a simple application that can be modeled with a Petri net.

3.6.1 Automata and Finite State Systems

Assume that we have a DES model

$$G_{aut} = (Q, \Sigma, \delta, \boldsymbol{E}),$$

where Q is the set of states, Σ is the set of events,

$$\delta : \Sigma \times Q \longrightarrow Q$$

is the state transition function, and we allow all event trajectories (denoted by $\boldsymbol{E}$) to occur (such a DES is often called a "logical" DES (logical DES are a class of discrete time asynchronous DES with equations of motion that are most often nonlinear and discontinuous with respect to the random occurrence of events). We emphasize that for G_{aut} we focus on general logical DES models where the state and event sets Q and Σ are nonnumeric, that is, "symbolic," and there are no particular assumptions about δ. In this general case, even though the "state space" of G_{aut} is completely unstructured, one can still metricize Q with the discrete metric ρ_d (where $\rho_d(q, q') = 0$ if $q \neq q'$ and $\rho_d(q, q') = 1$ if $q = q'$).

Relative to the metric space $\{Q; \rho_d\}$ any closed invariant set $Q_m \subset Q$ for G_{aut} is stable in the sense of Lyapunov w.r.t. $\boldsymbol{E}$ and asymptotically stable w.r.t. $\boldsymbol{E}$. This is the case since these are *local* properties. That is, for Lyapunov stability if you pick any $\epsilon > 0$ it is always possible to choose a $\delta > 0$ that is so small that it is not possible for *any* trajectory to start outside the invariant set; hence, the system is stable in the sense of Lyapunov since there are no trajectories that end up with a size bigger than ϵ since there are none possible.

For asymptotic stability in the large w.r.t. $\boldsymbol{E}$ we can let $V(q) = \rho_d(q, Q_m)$. Proving that $\rho_d(q_k, Q_m) \longrightarrow 0$ as $k \longrightarrow \infty$ for all possible initial states and event trajectories involves showing that for all possible event trajectories and initial states there exists $k' > 0$ such that $\rho_d(q_{k'}, Q_m) = 0$. Hence the Lyapunov framework for a metric space offers little in the way of analysis in such general cases (the analysis reduces to the study of invariant sets).

For finite state systems defined on a metric space it is the case that for all $x, x' \in \mathcal{X}$ there exists $\gamma > 0$ such that $\rho(x, x') > \gamma$. Hence, all G such that $|\mathcal{X}|$ is finite are stable in the sense of Lyapunov and asymptotically stable as in the automata model case. As for the Petri net case discussed below, the analysis of asymptotic stability in the large, exponential stability in the large, or the boundedness properties for finite state systems can, in some cases, be facilitated with the Lyapunov framework.

Next, we analyze the boundedness properties of Petri nets.

3.6.2 Petri Nets

Let $\xi = [\xi_1, \xi_2, \ldots, \xi_m]^\top$ such that $\xi \in \Re^m$ and $\xi_i > 0$, $i = 1, 2, \ldots, m$. Throughout this section we will use the metric $\rho : \mathbb{N}^m \times \mathbb{N}^m \to \Re$, where

$$\rho(M, M') = \sum_{i=1}^{m} \xi_i |M(p_i) - M'(p_i)| \tag{3.1}$$

and we will use $\mathcal{X}_b \subset \mathbb{N}^m$ to denote a bounded set. Next we state the standard definitions of boundedness for Petri nets [61, 88].

A Petri net $(\mathcal{N}, M_0)$ is said to be *γ-bounded* or simply *bounded* if for a given γ, $M(p_i) \leq \gamma$ for all $p_i \in P$ and $M \in R(M_0)$. A Petri net $\mathcal{N}$ is said to be *structurally bounded* if it is bounded for any finite initial marking M_0. For a Petri net $(\mathcal{N}, M_0)$: (1) $(\mathcal{N}, M_0)$ is γ-bounded for some $\gamma \geq 0$ iff the motions of $(\mathcal{N}, M_0)$ which begin at M_0 are bounded, (2) $\mathcal{N}$ is structurally bounded iff $\mathcal{N}$ possesses Lagrange stability, and (3) $\mathcal{N}$ is structurally bounded iff the motions of $\mathcal{N}$ are uniformly bounded. Next, we show how the Petri net-theoretic approach to the analysis of structural boundedness is actually a Lyapunov stability-theoretic approach. Moreover, we introduce the characterization and analysis of uniform ultimate boundedness for Petri nets.

Theorem 3.7 (Boundedness of Petri Nets): *For the Petri net $\mathcal{N}$ with $\mathcal{X}_b = \{0\}$:*

(*i*) *$\mathcal{N}$ is uniformly bounded if there exists an m-vector $\phi > 0$ such that $A\phi \leq 0$ and*

(*ii*) *$\mathcal{N}$ is uniformly ultimately bounded if there exists an m-vector $\phi > 0$ and n-vector $\pi > 0$ such that $A\phi \leq -\pi$.*

Proof: For (*i*) the proof follows by extending the one for structural boundedness in [61]. Let $\xi = \phi$ and choose

$$V(M) = \inf \left\{ \sum_{i=1}^{m} \phi_i |M(p_i) - M''(p_i)| : M'' \in \mathcal{X}_b \right\} = M^\top \phi, \tag{3.2}$$

so that due to the choice of ρ in Equation (3.1), the appropriate ψ_1 and ψ_2 exist so that

$$\psi_1(\rho(M, \mathcal{X}_b)) \leq V(M) \leq \psi_2(\rho(M, \mathcal{X}_b)).$$

Notice that V must only be defined and satisfy the appropriate properties on

$$\{M : \rho(M, \mathcal{X}_b) \geq R\},$$

where R may be large. Choose

$$R = \inf\{r' : 0 < \rho(M, \mathcal{X}_b) < r' \text{ and all } t \in T \text{ are enabled at } M\}$$

(R is finite since $W(p_i, t_j)$ is finite.) For (i), it suffices to show that for all M and $M' \in R_1(M)$ such that $M \in \{M : \rho(M, \mathcal{X}_b) \geq R\}$, $M'^{\top}\phi \leq M^{\top}\phi$. We know that for all M and $M' \in R_1(M)$, $M' = M + A^{\top}u$ for some $u \geq 0$ (we know that $u \geq 0$ exists since $M \in \{M : \rho(M, \mathcal{X}_b) \geq R\}$) and $M'^{\top} = M^{\top} + u^{\top}A$ so $M'^{\top}\phi = M^{\top}\phi + u^{\top}A\phi$. Since $u \geq 0$, $A\phi \leq 0$ implies that for all M and $M' \in R_1(M)$, $M'^{\top}\phi \leq M^{\top}\phi$ whenever $M \in \{M : \rho(M, \mathcal{X}_b) \geq R\}$.

For (ii), it suffices to show that for all M and $M' \in R_1(M)$ such that $M \in \{M : \rho(M, \mathcal{X}_b) \geq R\}$, $M'^{\top}\phi \leq M^{\top}\phi - \gamma$ for some $\gamma > 0$. From Equation (2.9), if $M' \in R_1(M)$, then $M'^{\top}\phi = M^{\top}\phi + u^{\top}A\phi$ so that $M'^{\top}\phi \leq M^{\top}\phi - u^{\top}\pi$. Since $u \geq 0$ exists (as long as $M \in \{M : \rho(M, \mathcal{X}_b) \geq R\}$) and $\pi_i > 0$,

$$M'^{\top}\phi - M^{\top}\phi \leq -\min\{\pi_i\}$$

for all M and $M' \in R_1(M)$. Hence, if we choose

$$\psi_3(\rho(M, \mathcal{X}_b)) = \min\{\pi_i\}\left(\frac{\rho(M, \mathcal{X}_b)}{1 + \rho(M, \mathcal{X}_b)}\right)$$

(ii) holds. ∎

Corollary 2: For the Petri net $\mathcal{N}$ with $\mathcal{X}_b = \{0\}$ if for each $t_j \in T$,

$$\sum_p W(p, t_j) \geq \sum_p W(t_j, p) \quad \left(\sum_p W(p, t_j) > \sum_p W(t_j, p)\right)$$

then $\mathcal{N}$ is uniformly bounded (resp., uniformly ultimately bounded).

Proof: Let $V(M) = M^{\top}\phi$, $\phi = [1, 1, \ldots, 1]^{\top}$, and use Theorem 3.7. ∎

An analogous result to Theorem 3.7 part (ii) exists for asymptotic stability in the large. Theorem 3.7 shows that the standard approach to boundedness analysis for general Petri nets is actually a special case of a Lyapunov approach to boundedness analysis. Really what is shown is that in the Petri net-theoretic approach to the analysis of structural boundedness [61], in picking ϕ one is actually picking a Lyapunov function $V(M) = M^{\top}\phi$. Once this

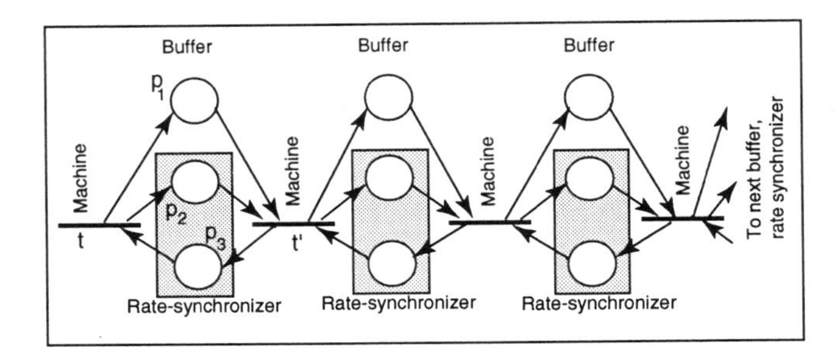

Fig. 3.1 Manufacturing line with rate-synchronization.

is recognized it will perhaps be easier to study boundedness properties due to the wealth of experience there is with regard to the choice of Lyapunov functions.

It is important to note that the Lyapunov approach also applies to the many subclasses of Petri nets (e.g., marked graphs and state machines).

3.6.3 Rate Synchronized Manufacturing Line

In this section we study boundedness properties of a special class of manufacturing lines with rate synchronization shown in Figure 3.1. We use a Petri net model for the manufacturing system and, hence, the results of the previous section for the analysis of the boundedness properties.

Suppose that in Figure 3.1 the transitions represent machines (a transition firing represents the completion of processing a part), and the places are used as shown to represent buffers where parts are passed through the system for processing (e.g., $M(p_1)$ represents the number of parts that have already been processed by the first machine and that are waiting to be processed by the second machine). The "rate-synchronizers" are used to ensure that the rates of processing of parts in the manufacturing system line are synchronized (to allow maximum flexibility in processing, we only seek to maintain a loosely coupled form of rate synchronization). Let $\mathcal{N} = (P, T, F, W)$ represent a manufacturing system with N such machines connected in series (similar analysis applies for other topologies). With this, for the Petri net framework we have $m = 3(N-1)$ and $n = N$.

For the analysis of boundedness properties choose $V(M) = M^\top \phi$ where $\phi = [1, 1, 2, 1, 1, 2, \ldots, 2]^\top$. Notice that if either transition t or t' in Figure 3.1 fires $V(M_{k+1}) \leq V(M)$ so that the manufacturing line with rate-synchronization is uniformly bounded. The choice of the "2" in the ϕ vector weights the adding and subtracting of tokens to, for example, place p_3, so that the weighted sum of tokens for the network will not increase. Checking that $A\phi \leq 0$ per part (i) of Theorem 3.7 also verifies the uniform boundedness of the manufacturing line.

3.7 SUMMARY

In this chapter we have provided mathematical definitions of a wide variety of stability concepts and have provided sufficient conditions to ensure that they hold. We have explained how to apply the Lyapunov framework to automata, finite state machines, and Petri nets and have provided a rate-synchronized manufacturing line as an example of how to perform analysis of boundedness properties.

In summary, the major topics covered in this chapter were:

- Stability concepts for the following:
  - Stability in the sense of Lyapunov.
  - Asymptotic stability.
  - Asymptotic stability in the large.
  - Exponential stability.
  - Exponential stability in the large.
  - Lagrange stability.
  - Uniform boundedness.
  - Uniform ultimate boundedness.
- Sufficient conditions for all of these stability concepts.
- Reachability and cyclic properties.
- Relevance of the stability framework to automata, finite-state machines, and Petri nets.
- Conditions for boundedness properties of Petri nets.
- Rate-synchronized manufacturing line boundedness analysis.

It is recommended that the reader clearly understand all of these topics before moving to later chapters, since all the later chapters involve applying the stability concepts and analysis to a variety of systems.

3.8 FOR FURTHER STUDY

It has been long known (as shown in, e.g., [110]) that a stability theory can be developed in a very broad setting (e.g., a metric space), which is phrased in terms of motions of dynamical systems and which does not require the description of the system under investigation in terms of specific equations (e.g., differential or difference equations, partial differential equations, etc.). The foundations for the study of stability properties of logical DES lie in the areas of general stability theory (the approach used herein) and theoretical Computer Science.

The metric space formulation for stability analysis in this chapter was developed in [78, 77, 81] (for stability in the sense of Lyapunov and asymptotic stability) and [75, 80] (for the boundedness properties and practical and finite time stability). The analysis of practical stability and finite time stability builds on the work in [54, 57]. For a detailed treatment of stability preserving mappings and comparison theory for systems defined on a metric space, see [58, 59, 101, 102] (there are DES applications there also—including an interconnected DES modeled with a Petri net). The fact that systems represented by Petri nets are amenable to Lyapunov stability analysis was first pointed out in [77, 81]. Other DES, that are similar to Petri nets, that can be studied in the metric space formulation here include [47, 46].

As mentioned above, Lyapunov concepts have been already been studied on a metric space (see, e.g., [110] for an introductory treatment, [53] for more advanced studies of stability preserving mappings on metric spaces and their applications, and [59] for more recent work on the use of a metric space Lyapunov approach for interconnected systems). There have also been studies of stability for more general topological spaces (see, e.g., [99]). There are many good books on conventional stability theory, including [35, 107, 56, 55].

The two (related) main areas in theoretical Computer Science that form the foundation for logical DES-theoretic stability studies are temporal logic and automata. Intuitively speaking, in a temporal logic or automata-theoretic framework a system is considered in some sense stable if (1) for some set of initial states the system's state is guaranteed to enter a given set and stay there forever, or (2) for some set of initial states the system's state is guaranteed to visit a given set of states infinitely often. In temporal logic, stability characteristics are most often represented with temporal formulas from a linear or branching time language (modal logics) and either a proof system or effective procedure is used to verify that the temporal formula is satisfied. The fact that the above notions of stability could be studied using temporal logic in a control-theoretic setting was first recognized in [29]. The linear time temporal logic framework of [52], which uses a proof system, is adapted and used to prove stability properties in a DES theoretic framework in [98]. A linear time temporal logic framework, where effective procedures are used to mechanically test the satisfaction of formulas describing stability properties, is studied in [40]. The branching time temporal logic approach in [22] is adapted to a DES theoretic framework and efficient algorithms are used to perform some studies of stability properties in [79]. Stability concepts for logical DES, such as finite automata, have foundations in the study of, for instance, Buchi and Muller automata [12, 60], and how infinite strings are accepted by such automata. This automata theoretic work in Computer Science has also been adapted for the study of stability of DES. In [68] the authors introduce a special DES model (finite automaton) and use a state-space approach to develop efficient algorithms for the study of two types of stability. They also provide approaches to synthesize stabilizing controllers for DES and to study several other characteristics of logical DES (for more details see [67]). A related studies is given in [9, 96].

Certain general formulations for the study of stability are relevant to the study of stability properties of logical DES. For instance, there have been studies of stability of asynchronous iterative processes in [99]. Tsitsiklis defines a model that can represent logical DES and, assuming that the DES has certain timing characteristics, he gives constructive methods to study stability of a class of DES. Tsitsiklis identifies the relationship between his work and the use of Lyapunov functions and provides some efficient procedures for testing stability. For an introduction to general stability theory and an overview of such research see [22].

3.9 PROBLEMS

Problem 3.1 (Stability Proofs): Prove the following two theorems.

Theorem 3.8 (Lyapunov Stability): *In order for a closed invariant set $\mathcal{X}_m \subset \mathcal{X}$ of G to be stable in the sense of Lyapunov with respect to $\boldsymbol{E}_a$, it is necessary and sufficient that in a sufficiently small neighborhood $S(\mathcal{X}_m; r)$ of the set $\mathcal{X}_m$ there exists a specified functional V with the following properties:*

(i) For all sufficiently small $c_1 > 0$, it is possible to find a $c_2 > 0$ such that $V(\boldsymbol{x}) > c_2$ for $\boldsymbol{x} \in S(\mathcal{X}_m; r)$ and $\rho(\boldsymbol{x}, \mathcal{X}_m) > c_1$.

(ii) For any $c_4 > 0$ as small as desired, it is possible to find a $c_3 > 0$ so small that when $\rho(\boldsymbol{x}, \mathcal{X}_m) < c_3$ for $\boldsymbol{x} \in S(\mathcal{X}_m; r)$ we have $V(\boldsymbol{x}) \leq c_4$.

(iii) $V(X(\boldsymbol{x}_0, E_k, k))$ is a nonincreasing function for $\boldsymbol{x}_0 \in S(\mathcal{X}_m; r)$ and for all $k \in N$, provided that $X(\boldsymbol{x}_0, E_k, k) \in S(\mathcal{X}_m; r)$ for all E_k, such that $E_k E \in \boldsymbol{E}_a(\boldsymbol{x}_0)$.

Theorem 3.9 (Asymptotic Stability): *In order for a closed invariant set $\mathcal{X}_m \subset \mathcal{X}$ of G to be asymptotically stable in the sense of Lyapunov with respect to $\boldsymbol{E}_a$, it is necessary and sufficient that, in a sufficiently small neighborhood, $S(\mathcal{X}_m; r)$, of the set $\mathcal{X}_m$ there exists a specified functional V having properties (i), (ii), and (iii) of Theorem 3.8 and furthermore $V(X(\boldsymbol{x}_0, E_k, k)) \longrightarrow 0$ as $k \longrightarrow \infty$ for all E_k, such that $E_k E \in \boldsymbol{E}_a(\boldsymbol{x}_0)$ and for all $k \in N$, as long as $X(\boldsymbol{x}_0, E_k, k) \in S(\mathcal{X}_m; r)$.*

Problem 3.2 (Exponential Stability Proofs): Prove the following theorems.

Theorem 3.10 (Exponential Stability: Another Approach): *In order for the invariant set $\mathcal{X}_m$ to be exponentially stable w.r.t. $\boldsymbol{E}_a$*

it is sufficient that in a neighborhood $S(\mathcal{X}_m; r)$ there exists a specified functional V and three positive constants c_1, c_2, and c_3 and

(i) $\psi_1(\rho(\boldsymbol{x}, \mathcal{X}_m)) \leq V(\boldsymbol{x}) \leq \psi_2(\rho(\boldsymbol{x}, \mathcal{X}_m))$, *and*

(ii) *we have*

$$V(X(\boldsymbol{x}_0, E_{k+1}, k+1)) - V(X(\boldsymbol{x}_0, E_k, k)) \leq \\ -c_3(\rho(X(\boldsymbol{x}_0, E_{k+1}, k+1), \mathcal{X}_m))$$

for all $\boldsymbol{x}_0 \in S(\mathcal{X}_m; r)$ and for all E_k such that $E_{k+1} = E_k e$ $(e \in \mathcal{E})$ and $E_{k+1}E \in \boldsymbol{E}_a(\boldsymbol{x}_0)$ and all $k \geq 0$.

Theorem 3.11 (Exp. Stability: Yet Another Approach): *The closed invariant set $\mathcal{X}_m \subset \mathcal{X}$ of G is exponentially stable with respect to $\boldsymbol{E}_a$ if there exists a functional V defined on $S(\mathcal{X}_m; r)$, $D \in \{1, 2, \ldots\}$ and $c_1, c_2, c_3 > 0$ with $c_3/c_2 \in (0,1)$ such that*

(i) $c_1\rho(\boldsymbol{x}, \mathcal{X}_m) \leq V(\boldsymbol{x}) \leq c_2\rho(\boldsymbol{x}, \mathcal{X}_m)$ *for all $\boldsymbol{x} \in S(\mathcal{X}_m; r)$*

(ii) *we have*

$$V(X(\boldsymbol{x}_0, E_{k+D}, k+D)) - V(X(\boldsymbol{x}_0, E_k, k)) \leq \\ -c_3\rho(X(\boldsymbol{x}_0, E_k, k), \mathcal{X}_m)$$

for all $\boldsymbol{x}_0 \in S(\mathcal{X}_m; r)$, $k \in N$, provided that $X(\boldsymbol{x}_0, E_k, k) \in S(\mathcal{X}_m; r)$ for all E_k, such that $E_k E \in \boldsymbol{E}_a(\boldsymbol{x}_0)$.

Problem 3.3 (Stability Analysis of a Simple Load Balancing System): In this problem you will analyze the properties of the load balancing system that is defined in Section 2.3.2 on page 12.

(a) Define an invariant set that represents the load being balanced (pay special attention to the fact that we consider discrete load and hence it may not be possible to perfectly balance the load – why?). Define a set of allowable event trajectories.

(b) Is this invariant set stable in the sense of Lyapunov? Show why.

(c) Is this invariant set asymptotically stable? Show why.

(d) Is this invariant set asymptotically stable in the large? Show why.

(e) Is this invariant set exponentially stable? Show why.

(f) Is this invariant set exponentially stable in the large? Show why.

Problem 3.4 (Boundedness Analysis of a One-Buffer Machine): In this problem you will analyze the boundedness properties of the single-buffer machine in Section 2.3.1 on page 11. Does the single buffer machine have a bounded buffer? If so, show that it does. If not, give an event sequence that will result in an unbounded buffer level.

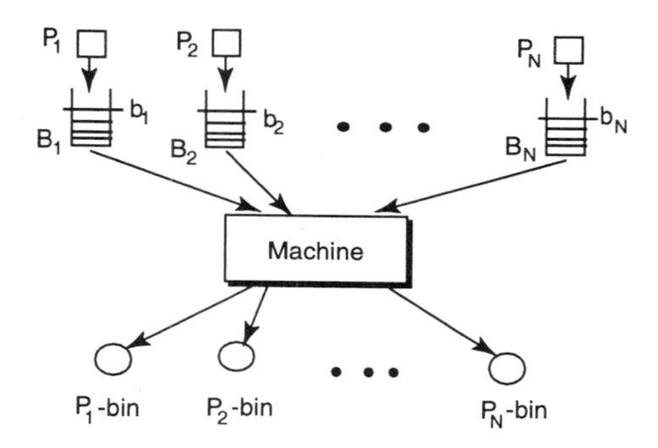

Fig. 3.2 Manufacturing system with priority batch processing.

Problem 3.5 (Stability Analysis of a Production Network): Characterize and analyze the stability properties of the production network in Section 2.5.1 on page 16. In particular, determine all stability properties that the production network has. Show all your reasoning and, if it possesses some property, then prove, using the Lyapunov framework, that it does.

Problem 3.6 (Stability Analysis of a Computer Network): Characterize and analyze the stability properties of the computer network in Section 2.5.2 on page 18. In particular, determine all stability properties that the computer network has. Show all your reasoning and, if it possesses some property, then prove, using the Lyapunov framework, that it does.

Problem 3.7 (Stability Analysis of a Manufacturing System)*: Consider the manufacturing system shown in Figure 3.2 that processes batches of N different types of jobs according to a priority scheme. Here we use the term "job" in a general sense. For us, the completion of a job may mean the processing of a batch of 10 parts, the processing of a batch of 5.103 tasks, etc. There are N producers P_i, where $1 \leq i \leq N$, of jobs of different types. The producers place batches of their jobs in their respective buffers B_i, where $1 \leq i \leq N$. These buffers B_i have safe capacity limits of b_i where $b_i > 0$, $1 \leq i \leq N$. Let x_i, $1 \leq i \leq N$, denote the number of jobs in buffer B_i. Let x_i for $N+1 \leq i \leq 2N$ denote the number of P_{i-N} type jobs in the machine. The machine can safely process less than or equal to M (where $M > 0$) jobs of any type, at any time. As the machine finishes processing batches of P_i type jobs, they are placed in their respective output bins (P_i bins). The producers P_i can only place batches of jobs in their buffers B_i if $x_i < b_i$. Also, there is a priority scheme whereby batches of P_i type jobs are only allowed to enter the machine if $x_j = 0$ for all j, such that $j < i \leq N$.

(a) Specify the DES model G for the manufacturing system.

(b) Specify a Petri net model for the manufacturing system.

(c) Specify a metric and an invariant set.

(d) Determine all stability properties that the manufacturing system has. Show all your reasoning and, if it possesses some property, then prove, using the Lyapunov framework, that it does.

4

Load Balancing in Computer Networks

4.1 OVERVIEW

A computer that is used as a *load processor* has a buffer that can receive load (e.g., tasks or jobs) and store it while it is waiting to be processed and has a local decision-making policy for determining if portions of its load should be sent to other load processors. A *load balancing system* is a set of such load processors that are connected in a network so that they can sense the amount of load in the buffers of neighboring processors and pass load to them, and so that, via local information and decisions by the individual load processors, the overall load in the entire network can be balanced. Such balancing is important to ensure that certain processors are not overloaded while others are left idle (i.e., load balancing helps avoid under-utilization of processing resources).

While the load balancing problem we study here originated in the computer networks area, there may be a need for load balancing in other applications. For example, it is easy to envision the need for balancing the number of parts to be processed in a manufacturing system to avoid under-utilization of a machine. There are also applications in a variety of other queueing systems.

The topology of the load balancing network, delays in transporting and sensing load, types of load, and types of local load passing policies all affect the performance and operation of the load balancing system. In this chapter, we show how a variety of load balancing systems can be modeled with the DES model in Section 2.2, and how balancing properties and performance can be characterized and analyzed in the general Lyapunov stability theoretic framework of Chapter 3. Our main objective in this chapter is to show how to analyze:

- Stability in the sense of Lyapunov
- Asymptotic stability
- Asymptotic stability in the large
- Exponential stability
- Exponential stability in the large

for a complex application. We view this chapter as a case study in stability analysis of the above-mentioned properties. In Chapter 5 we perform another case study in stability analysis, but for the boundedness properties in Chapter 3. Finally, note that, while this chapter will help to solidify some concepts and analysis approaches in Chapter 3, Chapters 5 and 6 depend in no way on your understanding of this chapter.

4.2 A LOAD BALANCING PROBLEM WITHOUT DELAYS

A load balancing system is a network of load processors (e.g., machines in a manufacturing system, computers on a network) that are connected together so that any processor on the network is capable of passing a portion of its load (e.g., jobs, tasks, parts) to any other processor to which it is connected and, if a processor can pass load to another load processor, it can also sense the load level of that processor. Figure 4.1 illustrates an example load balancing system, where each load processor, along with its buffer, is numbered from 1 to 6 and the arc from 1 to 2 indicates that 1 can sense the amount of load in the buffer of processor 2 and that 1 can pass load to processor 2. Since there are no arcs between 1 and 5, these processors cannot sense each others' loads or pass load to each other. Next, we will consider a general network topology.

The load processors, $L = \{1, 2, ..., N\}$, are all connected to a network along which they can pass load to other load processors. The network of load processors is described by a directed graph, (L, A), where $A \subset L \times L$. For every $i \in L$, there must exist $(i, j) \in A$ in order to assure that every load processor is connected to the network, and if $(i, j) \in A$ then $(j, i) \in A$. Load processor i can only transfer a portion of its load to load processor j if $(i, j) \in A$. Finally, if $(i, j) \in A$, then $i \neq j$.

Each load processor $i \in L$ has a buffer in which its load is stored prior to processing. It is the buffer levels x_i that we actually wish to balance; thus, it is the buffer levels that are affected by load transfers. In this section, we will assume that the load can be partitioned into sufficiently small units so that it is valid to describe it with a continuous variable. We will also assume that

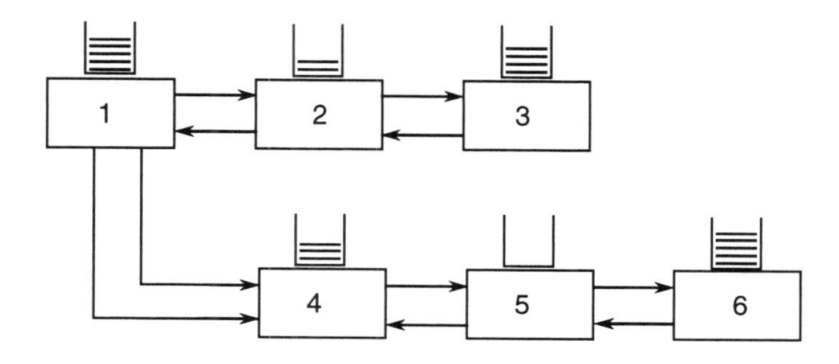

Fig. 4.1 Example of a load balancing system.

the total amount of load in the buffers of the load processors on the network remains static until a load balance is achieved; hence we assume that no load arrives or is processed during the balancing of the load.

In this section, we are not considering load transportation or load sensing delays. Hence, we require that the *real time* between events e_k and e_{k+1} (which will represent the passing of load) be greater than the greatest system transportation time plus the greatest system sensing time. We do, however, allow for more than one node to pass load at one time and for nodes to simultaneously pass load to more than one of the nodes that they are connected to on the network.

We begin by specifying the DES model G. Let $\mathcal{X} = \Re^N$ be the set of states and $\boldsymbol{x}_k = [x_1, x_2, ..., x_N]^\top$ and $\boldsymbol{x}_{k+1} = [x'_1, x'_2, ..., x'_N]^\top$ denote the states at times k and $k+1$, respectively. Let $x_i(k')$ denote the amount of load at node $i \in L$ at time k'. Let $e^{i,p(i)}_{\alpha(i)}$ represent that node $i \in L$ passes load to its *neighbors* $m \in p(i)$, where $p(i) = \{j : (i,j) \in A\}$. Let the list $\alpha(i) = (\alpha_j(i), \alpha_{j'}(i), \ldots, \alpha_{j''}(i))$ such that $j < j' < \cdots < j''$ and $j, j', \ldots, j'' \in p(i)$ and $\alpha_j \geq 0$ for all $j \in p(i)$; the size of the list $\alpha(i)$ is $|p(i)|$. For convenience, we will denote this list by $\alpha(i) = (\alpha_j(i) : j \in p(i))$. $\alpha_m(i)$ denotes the amount of load transferred from $i \in L$ to $m \in p(i)$. Let $\{e^{i,p(i)}_{\alpha(i)}\}$ denote the set of all possible such load transfers. Let the set of events be described by

$$\mathcal{E} = \mathcal{P}(\{e^{i,p(i)}_{\alpha(i)}\}) - \{\phi\},$$

(recall that $\mathcal{P}(Q)$ denotes the power set of the set Q). Notice that each event $e_k \in \mathcal{E}$ is defined as a set, with each element of e_k representing the passing of load by some node $i \in L$ to its neighboring nodes in the network. Let $\gamma_{ij} \in (0,1)$ for $(i,j) \in A$ represent the proportion of the load imbalance that is *sometimes* guaranteed to be reduced when i passes load to j.

Below, we specify g and f_{e_k} for $e_k \in g(\boldsymbol{x}_k)$:

- Event $e_k \in g(\boldsymbol{x}_k)$ if both (a) and (b) below hold:

(a) For all $e^{i,p(i)}_{\alpha(i)} \in e_k$, where $\alpha(i) = (\alpha_j(i) : j \in p(i))$ it is the case that:

$$
\begin{array}{ll}
(i) & \alpha_j(i) = 0 \text{ if } x_i \leq x_j, \text{ where } j \in p(i), \\
(ii) & 0 \leq \sum_{m \in p(i)} \alpha_m(i) \leq x_i - (x_j + \alpha_j(i)), \text{ for all } j \in p(i), \\
 & \text{such that } x_i > x_j, \text{ and} \\
(iii) & \alpha_{j^*}(i) \geq \gamma_{ij^*}(x_i - x_{j^*}) \text{ for some} \\
 & j^* \in \{j : x_j \leq x_m, \text{ for all } m \in p(i)\}.
\end{array}
$$

Condition (i) prevents load from being passed by node i to node j if node i is less heavily loaded than node j. Condition (ii) directly implies that $x_i - \sum_{m \in p(i)} \alpha_m(i) \geq x_j + \alpha_j(i)$. Thus, after the load $\alpha(i)$ has been passed, the remaining load of node i must be at least as large as $x_j + \alpha_j(i)$ for every node $j \in p(i)$ that was less heavily loaded than node i to begin with. Condition (iii) implies that if node i is not load balanced with all of its neighbors and it passes load, then i must pass a nonnegligible portion of its load to some least loaded neighbor j^*.

(b) If $e^{i,p(i)}_{\alpha(i)} \in e_k$, where $\alpha(i) = (\alpha_j(i) : j \in p(i))$, then $e^{i,p(i)}_{\delta(i)} \notin e_k$, where $\delta(i) = (\delta_j(i) : j \in p(i))$ if $\alpha_j(i) \neq \delta_j(i)$ for some $j \in p(i)$. Hence, in each valid event e_k, there must be a consistent definition of the load to be passed from any node i to any other node j, $\alpha_j(i)$.

- If $e_k \in g(\boldsymbol{x}_k)$ and $e^{i,p(i)}_{\alpha(i)} \in e_k$, then $f_{e_k}(\boldsymbol{x}_k) = \boldsymbol{x}_{k+1}$, where

$$
x'_i = x_i - \sum_{\{j:\, j \in p(i)\}} \alpha_j(i) + \sum_{\{j:\, i \in p(j)\,,\, e^{j,p(j)}_{\alpha(j)} \in e_k\}} \alpha_i(j).
$$

The load of node i at time $k+1$, x'_i, is the load of node i at time k minus the total load passed by node i at time k plus the total load received by node i at time k.

Let $\boldsymbol{E}_v = \boldsymbol{E}$ be the set of valid event trajectories. We must further specify the sets of allowed event trajectories. Define a *partial event of type* i to represent the passing of $\alpha(i)$ amount of load from $i \in L$ to its neighbors $p(i)$. A partial event of type i will be denoted by $e^{i,p(i)}$ and the occurrence of $e^{i,p(i)}$ indicates that $i \in L$ attempts to further balance its load with its neighbors. Event $e_k \in g(\boldsymbol{x}_k)$ is composed of a set of partial events. Next we define two possibilities for the allowed event trajectories $\boldsymbol{E}_a$.

(i) For $\boldsymbol{E}_i \subset \boldsymbol{E}_v$, assume that each type of partial event occurs infinitely often on each $E \in \boldsymbol{E}_i$.

(*ii*) For $\boldsymbol{E}_B \subset \boldsymbol{E}_v$, assume that there exists $B > 0$, such that for every event trajectory $E \in \boldsymbol{E}_B$, in every substring $e_{k'}, e_{k'+1}, e_{k'+2}, \ldots, e_{k'+(B-1)}$ of E there is the occurrence of every type of partial event (i.e. for every $i \in L$ partial event $e^{i,p(i)} \in e_k$, for some k, $k' \leq k \leq k' + B - 1$).

Clearly,

$$\mathcal{X}_b = \{\boldsymbol{x_k} \in \mathcal{X} : x_i = x_j, \text{ for all } (i,j) \in A\}$$

is an invariant set that represents a perfectly balanced load (not to be confused with $\mathcal{X}_b$ from Chapter 3). Notice that the only $e_k \in g(\boldsymbol{x}_k)$, when $\boldsymbol{x}_k \in \mathcal{X}_b$, are ones such that all $e^{i,p(i)}_{\alpha(i)} \in e_k$ have $\alpha(i) = (0, 0, \ldots, 0)$.

If $\boldsymbol{E}_a = \boldsymbol{E}_B \subset \boldsymbol{E}_i$, the load balancing problem described above is the same as the one in [100], except that in this section we do not allow delays in transporting and sensing load. In Section 4.4 we will study load balancing systems with delays.

4.2.1 Asymptotic Convergence to a Balanced State

To study the ability of the system to automatically redistribute load to achieve balancing, we use a Lyapunov stability theoretic approach. Let $\bar{\boldsymbol{x}} = [\bar{x}_1, ..., \bar{x}_N]$. Choose

$$\rho(\boldsymbol{x}_k, \mathcal{X}_b) = \inf\{\max\{|x_1 - \bar{x}_1|, ..., |x_N - \bar{x}_N|\} : \bar{\boldsymbol{x}} \in \mathcal{X}_b\}. \tag{4.1}$$

Theorem 4.1 (Continuous Load, No Delays, Asymptotic Convergence): *For the load processor network system described above, the invariant set $\mathcal{X}_b$ is asymptotically stable in the large with respect to $\boldsymbol{E}_i$.*

Proof: Choose

$$V(\boldsymbol{x}_k) = \max_i \left\{ \frac{1}{N} \sum_{j=1}^{N} x_j - x_i \right\}. \tag{4.2}$$

Notice that

$$\frac{1}{N} \sum_{j=1}^{N} x_j \geq \frac{1}{N} [\max_i\{x_i\} + (N-1) \min_i\{x_i\}]. \tag{4.3}$$

It is clear from Equations (4.1), (4.2), and (4.3) that the following relations are valid.

$$\rho(\boldsymbol{x}_k, \mathcal{X}_b) \geq \frac{1}{2} \left(\max_i\{x_i\} - \min_i\{x_i\} \right) \tag{4.4}$$

$$\rho(\boldsymbol{x}_k, \mathcal{X}_b) \leq \max_i\{x_i\} - \min_i\{x_i\} \tag{4.5}$$

$$V(\boldsymbol{x}_k) = \frac{1}{N} \sum_{j=1}^{N} x_j - \min_i\{x_i\} \leq \max_i\{x_i\} - \min_i\{x_i\} \tag{4.6}$$

$$V(\boldsymbol{x}_k) \geq \frac{1}{N} [\max_i\{x_i\} + (N-1) \min_i\{x_i\}] - \min_i\{x_i\} \tag{4.7}$$

Equations (4.4) and (4.6) yield $2\rho(\boldsymbol{x}_k, \mathcal{X}_b) \geq \max_i\{x_i\} - \min_i\{x_i\} \geq V(\boldsymbol{x}_k)$, so that condition (*ii*) of Theorem 3.8 is satisfied. Equation (4.7) can be manipulated to yield

$$V(\boldsymbol{x}_k) \geq \frac{1}{N}(\max_i\{x_i\} - \min_i\{x_i\}) \tag{4.8}$$

Equations (4.5) and (4.8) directly imply that

$$V(\boldsymbol{x}_k) \geq \frac{1}{N}\rho(\boldsymbol{x}_k, \mathcal{X}_b),$$

so that condition (*i*) of Theorem 3.8 is satisfied. Note that conditions (*i*) and (*ii*) of Theorem 3.8 are equivalent to condition (*i*) of Theorem 3.1.

For the final condition of Theorem 3.1, we must show that $V(X(\boldsymbol{x}_0, E_k, k))$ is a nonincreasing function for all $k \in$ N, all $\boldsymbol{x}_0 \in S(\mathcal{X}_b; r)$ and all E_k, such that $E_k E \in \boldsymbol{E}_i(\boldsymbol{x}_0)$. To see that this is the case, notice that once $\boldsymbol{x}_0$ is specified, $V(\boldsymbol{x}_k)$ varies only as the lightest load in the network, $\min_i\{x_i\} = x_{j^{**}}$ varies. The most lightly loaded node in the network cannot possibly pass load, so $x'_{j^{**}} \geq x_{j^{**}}$. Assume an event $e_k \in g(\boldsymbol{x}_k)$ occurs. According to condition (*ii*) on $e_k \in g(\boldsymbol{x}_k)$, if $e^{i,p(i)}_{\alpha(i)} \in e_k$ and $j^{**} \in p(i)$, it is not possible that $x'_i < x_{j^{**}} + \alpha_{j^{**}}(i)$. Therefore, $\min_i\{x'_i\} \geq x_{j^{**}}$ and $V(\boldsymbol{x}_{k+1}) \leq V(\boldsymbol{x}_k)$. Thus, condition (*ii*) of Theorem 3.1 is satisfied and $\mathcal{X}_b$ is stable in the sense of Lyapunov with respect to $\boldsymbol{E}_i$.

In order to show that $\mathcal{X}_b$ is asymptotically stable in the large with respect to $\boldsymbol{E}_i$, we must show that for all $\boldsymbol{x}_0 \notin \mathcal{X}_b$ and all E_k such that $E_k E \in \boldsymbol{E}_i(\boldsymbol{x}_0)$, $V(X(\boldsymbol{x}_0, E_k, k)) \longrightarrow 0$ as $k \longrightarrow \infty$. If $\boldsymbol{x}_k \notin \mathcal{X}_b$, then there must be some lightest loaded node j^{**} (there may be more than one such node) and some other node i such that $(i, j^{**}) \in A$ and $x_i > x_{j^{**}}$. Because of the restrictions imposed by $\boldsymbol{E}_i$, we know that all the partial events are guaranteed to occur infinitely often. According to condition (*a*)(*iii*) on $e_k \in g(\boldsymbol{x}_k)$, each time partial event $e^{i,p(i)}$ occurs, $x_{j^{**}}$ is guaranteed to increase by a fixed fraction $\gamma_{ij^{**}} \in (0,1)$ of $x_i - x_{j^{**}}$ so that $x'_{j^{**}} > x_{j^{**}}$. Thus, regardless of how many lightest loaded nodes there are, it is inevitable that eventually the overall lightest load in the network must increase. Hence, for every $k \geq 0$, there exists $k' > k$ such that $V(\boldsymbol{x}_{k'}) > V(\boldsymbol{x}_{k'+1})$ as long as $\boldsymbol{x}_{k'} \notin \mathcal{X}_b$ so that $V(X(\boldsymbol{x}_0, E_k, k)) \longrightarrow 0$ as $k \longrightarrow \infty$ and $\mathcal{X}_b$ is asymptotically stable in the large with respect to $\boldsymbol{E}_i$. ■

Consider the following issues:

- Notice that we do not need the restrictions on allowed event trajectories that are imposed by $\boldsymbol{E}_i$ to support our conclusion of stability in the sense of Lyapunov. Hence, $\mathcal{X}_b$ is stable in the sense of Lyapunov with respect to $\boldsymbol{E}_v$ as well.

- Note that $\mathcal{X}_b$ is not asymptotically stable in the large with respect to $\boldsymbol{E}_v$. This is due to the fact that without the restrictions on $\boldsymbol{E}_v$ to obtain

$\boldsymbol{E}_i$, it is possible that only one $i \in L$ attempts to balance its load for all time.

- Notice that condition $(a)(i)$ on $e_k \in g(\boldsymbol{x}_k)$ is absolutely necessary. If condition (i) is removed, then it is possible that nodes may pass load to their more heavily loaded neighbors. In this case, node j^{**} (where $x_{j^{**}} = \min_m\{x_m : m \in L\}$) may pass load and $x'_{j^{**}} < x_{j^{**}}$. Hence, the lightest load in the network may decrease and both the proof of Lyapunov stability and the proof of asymptotic stability become invalid.

- Consider the implications of replacing condition $(a)(ii)$ on $e_k \in g(\boldsymbol{x}_k)$ with the more liberal condition

$$0 \leq \sum_{j \in p(i)} \alpha_j(i) \leq x_i - x_j \quad \text{for all } \ j \in p(i), \quad \text{such that } \ x_i \geq x_j.$$

This new condition implies that if $e_{\alpha(i)}^{i,p(i)} \in e_k$, $\alpha_{j^*}(i)$ (where $j^* \in \{j : x_j \leq x_m \ \text{ for all } \ m \in p(i)\}$) may be such that $x'_i = x_{j^*}$ and $x'_{j^*} = x_i$. In this case, nodes i and j^* simply exchange load levels. It is still true that the lightest load in the network cannot decrease, however, it is not necessarily true that the lightest load in the network will ever increase. Hence, $\mathcal{X}_b$ remains stable in the sense of Lyapunov with respect to $\boldsymbol{E}_i$, but we can no longer claim that $\mathcal{X}_b$ is asymptotically stable with respect to $\boldsymbol{E}_i$.

- Consider eliminating condition $(a)(iii)$ on $e_k \in g(\boldsymbol{x}_k)$. In this case, if $e_{\alpha(i)}^{i,p(i)} \in e_k$ and $j^{**} \in p(i)$ (where $x_{j^{**}} = \min_m\{x_m : m \in L\}$), it is no longer true that $x_{j^{**}}$ must increase by a fixed fraction of $x_i - x_{j^{**}}$. It is now possible that even if $e_{\alpha(i)}^{i,p(i)} \in e_k$, for all $k > k'$, $x_{j^{**}} \not\longrightarrow x_i$ as $k \longrightarrow \infty$. For example, $x_i - x_{j^{**}}$ may be reduced after each load passing by factors of $1/(k+1)^2$ and the two loads will never converge to each other. Hence, it is no longer true that $\mathcal{X}_b$ is asymptotically stable with respect to $\boldsymbol{E}_i$, but it is still the case that $\mathcal{X}_b$ is stable in the sense of Lyapunov with respect to $\boldsymbol{E}_i$.

4.2.2 Exponential Convergence to a Balanced State

We now say something about the *rate* at which the system converges to a balanced state. In order to do this, we employ Theorem 3.11. If we satisfy the conditions of this theorem, we know that $\rho(\boldsymbol{x}_k, \mathcal{X}_b)$ will be bounded from above by an exponential $\zeta e^{-\alpha k} \rho(\boldsymbol{x}_0, \mathcal{X}_b)$ for some $\alpha > 0$ and $\zeta > 0$.

Theorem 4.2 (Continuous Load, No Delays, Exponential Stability): *For the load processor network system described above, the invariant set $\mathcal{X}_b$ is exponentially stable in the large with respect to $\boldsymbol{E}_B$.*

Proof: Choose the same $V(\boldsymbol{x})$ as in Equation (4.2). The first condition of the Theorem 3.11 is shown to hold in the proof of asymptotic stability, Theorem 4.1. We now show that the second condition of Theorem 3.11 holds.

Let $\gamma = \min_{i,j}\{\gamma_{ij}\}$. For any $i \in L$ and $k \geq 0$, we know from condition $(a)(ii)$ on $e_{\alpha(i)}^{i,p(i)} \in e_k$ and the definition of γ that if $e_{\alpha(i)}^{i,p(i)} \in e_k$ and $\alpha_j(i) > 0$ for some $j \in p(i)$, then $x_i' \geq x_j + \gamma(x_i - x_j)$ for some $j \in p(i)$. If $e_{\alpha(i)}^{i,p(i)} \notin e_k$ or $e_{\alpha(i)}^{i,p(i)} \in e_k$ and $\alpha(i) = (0, 0, \ldots, 0)$, then $x_i' = x_i$. It follows that in any case,

$$x_i' \geq \min_i\{x_i\} + \gamma[x_i - \min_i\{x_i\}]. \tag{4.9}$$

Thus, it is clear that $\min_i\{x_i\}$ is a nondecreasing function of k. We now show via induction on t that

$$x_i(k+t) \geq \min_i\{x_i\} + \gamma^t[x_i - \min_i\{x_i\}], \tag{4.10}$$

for all $t \geq 0$. Equation (4.9) is the statement of Equation (4.10) for $t = 1$. Assume that Equation (4.10) is true for an arbitrary t. If x_i denotes the load of $i \in L$ at time k, then according to Equation (4.9),

$$\begin{aligned} x_i(k+t+1) &\geq \min_i\{x_i(k+t)\} + \gamma[x_i(k+t) - \min_i\{x_i(k+t)\}] \\ &\geq \min_i\{x_i\} + \gamma[x_i(k+t) - \min_i\{x_i\}] \\ &\geq \min_i\{x_i\} + \gamma\left[\min_i\{x_i\} + \gamma^t[x_i - \min_i\{x_i\}] - \min_i\{x_i\}\right] \\ &= \min_i\{x_i\} + \gamma^{t+1}[x_i - \min_i\{x_i\}]. \end{aligned}$$

Thus, Equation (4.10) must be valid for all $t \geq 0$.

Fix $i \in L$ and $k \geq 0$. We now show that the loads of all neighbors of i are bounded from below by a function of x_i for all k', $k' \geq k + NB$. Specifically, we will show that

$$x_j(k') \geq \min_i\{x_i\} + \gamma^{k'-k}[x_i - \min_i\{x_i\}] \text{ for all } k' \geq k + NB\ ,\ \ j \in p(i). \tag{4.11}$$

There are times $k_m \geq k$, $m \in \{1, 2, \ldots\}$, such that $e_{\alpha(i)}^{i,p(i)} \in e_{k_m}$, and for $k' \neq k_m$, $e_{\alpha(i)}^{i,p(i)} \notin e_{k'}$. According to the restriction on $E \in \boldsymbol{E}_B$, $k \leq k_1 < k+B$ and $k_{m-1} < k_m < k_{m-1}+B$ for all $m \in \{2, 3, \ldots\}$. Below we investigate three cases that may occur at any time k_m. The different cases describe different possible relative load levels of node i and its neighbors. More than one case may apply to a given time k_m.

In the first case, there is time k_m, $m \in \{1, 2, \ldots\}$, and $j \in p(i)$ such that $x_j(k_m) < x_i(k_m)$ and $x_j(k_m) \leq x_{j'}(k_m)$ for all $j' \in p(i)$. According to condition $(a)(iii)$ on $e_k \in g(\boldsymbol{x}_k)$, $\alpha_j(i) \geq \gamma[x_i(k_m) - x_j(k_m)]$. Utilizing this fact and applying Equation (4.10) to x_i yields

$$\begin{aligned}
x_j(k_m+1) &\geq x_j(k_m) + \gamma[x_i(k_m) - x_j(k_m)] \\
&\geq \min_i\{x_i\} + \gamma[x_i(k_m) - \min_i\{x_i\}] \\
&\geq \min_i\{x_i\} + \gamma[\min_i\{x_i\} + \gamma^{k_m-k}[x_i - \min_i\{x_i\}] - \min_i\{x_i\}] \\
&= \min_i\{x_i\} + \gamma^{k_m-k+1}[x_i - \min_i\{x_i\}].
\end{aligned}$$

If we now apply Equation (4.10) to x_j with $k = k_m + 1$ and $t = k' - k_m - 1$, it is clear that

$$\begin{aligned}
x_j(k') &\geq \min_i\{x_i(k_m+1)\} + \gamma^{k'-k_m-1}[x_j(k_m+1) - \min_i\{x_i(k_m+1)\}] \\
&\geq \min_i\{x_i\} \\
&+ \gamma^{k'-k_m-1}[\min_i\{x_i\} + \gamma^{k_m-k+1}[x_i - \min_i\{x_i\}] - \min_i\{x_i\}] \\
&\geq \min_i\{x_i\} + \gamma^{k'-k}[x_i - \min_i\{x_i\}] \text{ for all } k' \geq k_m + 1. \qquad (4.12)
\end{aligned}$$

In the second case, there is time k_m, $m \in \{1, 2, \ldots\}$, and $j' \in p(i)$, such that at some time k_q, $1 \leq q < m$, $\alpha_{j'}(i) \geq \gamma[x_i(k_q) - x_{j'}(k_q)]$. In other words, at time k_q, node i passed at least $\gamma[x_i(k_q) - x_{j'}(k_q)]$ to node j'. We consider any $j \in p(i)$, such that $x_j(k_m) \geq x_{j'}(k_m)$. Applying Equation (4.10) to x_j with $k = k_m$ and $t = k' - k_m$ yields

$$\begin{aligned}
x_j(k') &\geq \min_i\{x_i(k_m)\} + \gamma^{k'-k_m}[x_j(k_m) - \min_i\{x_i(k_m)\}] \\
&\geq \min_i\{x_i\} + \gamma^{k'-k_m}[x_{j'}(k_m) - \min_i\{x_i\}] \qquad (4.13)
\end{aligned}$$

for all $k' \geq k_m$. Clearly, Equation (4.12) applies to node j' for all k', $k' \geq k_q + 1$. Because $k_m \geq k_q + 1$, we can substitute in Equation (4.13) for $x_{j'}(k_m)$ from Equation (4.12) to arrive at

$$\begin{aligned}
x_j(k') &\geq \min_i\{x_i\} + \gamma^{k'-k_m}[\min_i\{x_i\} + \gamma^{k_m-k}[x_i - \min_i\{x_i\}] - \min_i\{x_i\}] \\
&\geq \min_i\{x_i\} + \gamma^{k'-k}[x_i - \min_i\{x_i\}] \text{ for all } k' \geq k_m. \qquad (4.14)
\end{aligned}$$

In the third case, there is time k_m, $m \in \{1, 2, \ldots\}$, such that $x_i(k_m) \leq x_j(k_m)$, for all $j \in p(i)$ (i.e., all neighbors of node i are at least as heavily loaded as node i). In this case, for any $j \in p(i)$, it is clear from Equation (4.10) with $k = k_m$ and $t = k' - k_m$ that

$$\begin{aligned}
x_j(k') &\geq \min_i\{x_i(k_m)\} + \gamma^{k'-k_m}[x_j(k_m) - \min_i\{x_i(k_m)\}] \\
&\geq \min_i\{x_i\} + \gamma^{k'-k_m}[x_i(k_m) - \min_i\{x_i\}] \qquad (4.15)
\end{aligned}$$

for all $k' \geq k_m$. From Equation (4.10) with $t = k_m - k$, it is also clear that

$$x_i(k_m) \geq \min_i\{x_i\} + \gamma^{k_m-k}[x_i - \min_i\{x_i\}]. \qquad (4.16)$$

If follows then from Equations (4.15) and (4.16) that

$$\begin{aligned} x_j(k') &\geq \min_i\{x_i\} + \gamma^{k'-k_m}[\min_i\{x_i\} + \gamma^{k_m-k}[x_i - \min_i\{x_i\}] - \min_i\{x_i\}] \\ &\geq \min_i\{x_i\} + \gamma^{k'-k}[x_i - \min_i\{x_i\}] \text{ for all } k' \geq k_m. \end{aligned} \quad (4.17)$$

Now notice that at each time k_m, $m \in \{1, 2, \ldots\}$, it must be the case that exactly one of the following is true:

(*i*) There is at least one $j \in p(i)$ such that $\alpha_j(i) \geq \gamma[x_i(k_m) - x_j(k_m)]$ and at every time k_q, $q < m$, $\alpha_j(i) < \gamma[x_i(k_q) - x_j(k_q)]$ (i.e., Node i passes a nonnegligible amount of load at time k to at least one of its neighbors to which it has not passed a nonnegligible amount of load since before time k_1).

(*ii*) For every $j \in p(i)$ such that $\alpha_j(i) \geq \gamma[x_i(k_m) - x_j(k_m)]$, there is some $q < m$, such that the load passed by processor i to processor j at time q satisfies $\alpha_j(i) \geq \gamma[x_i(k_q) - x_j(k_q)]$ (i.e., Processor i passes a nonnegligible amount of load only to neighbors $j \in p(i)$ to which it has passed a nonnegligible amount of load since time k_1).

(*iii*) For every $j \in p(i)$, $x_i(k_m) \leq x_j(k_m)$ (i.e., Processor i cannot pass load to any of its neighbors $j \in p(i)$).

If (*ii*) is true, then the second case applies to all neighbors of i and Equation (4.14) is valid for all $j \in p(i)$. Hence, because $k_N \leq k + NB$, if $m < N$, then Equation (4.11) is valid. If (*iii*) is true, then the third case applies for all of the neighbors of i, and Equation (4.17) is valid for all $j \in p(i)$. Hence, because $k_N < k + NB$, if $m < N$, then Equation (4.11) is valid. If (*i*) is true, the first case applies to all of the neighbors of i to which i passes a nonnegligible amount of load, and Equation (4.12) is valid for all $j \in p(i)$ for which $\alpha_j(i) \geq \gamma[x_i(k_m) - x_j(k_m)]$ is true. Because $|p(i)| < N$, either (*ii*) or (*iii*) must occur before k_N or (*i*) must occur for every k_m, $m \in \{1, 2, \ldots, N-1\}$. Therefore, Equation (4.11) must be valid.

We now extend Equation (4.11) to

$$x_j(k') \geq \min_i\{x_i\} + (\gamma^{k'-k})^l[x_i - \min_i\{x_i\}] \text{ for all } k' \geq k + lNB, \quad (4.18)$$

where j is any node that is reachable from i by spanning l interprocessor connections (arcs $(i, j) \in A$). Equation (4.11) establishes the validity of Equation (4.18) for $l = 1$. We assume Equation (4.18) is valid for a general j at a distance l from i, and there must be some node $q \in p(j)$, such that q is at a distance $l + 1$ from i. Equation (4.11), applied to $q \in p(j)$, yields

$$\begin{aligned} x_q(k') \geq\ & \min_i\{x_i(k + lNB)\} + \\ & \gamma^{k'-(k+lNB)}[x_j(k + lNB) - \min_i\{x_i(k + lNB)\}] \end{aligned}$$

$$\begin{aligned} x_q(k') &\geq \min_i\{x_i\} + \gamma^{k'-k}[x_j(k+lNB) - \min_i\{x_i\}] \\ &\quad \text{for all } k' \geq k + (l+1)NB. \end{aligned}$$

Substituting, based on our inductive hypothesis,

$$\begin{aligned} x_q(k') &\geq \min_i\{x_i\} + \gamma^{k'-k}[\min_i\{x_i\} + (\gamma^{k'-k})^l[x_i - \min_i\{x_i\}] - \min_i\{x_i\}] \\ &= \min_i\{x_i\} + (\gamma^{k'-k})^{l+1}[x_i - \min_i\{x_i\}] \text{ for all } k' \geq k + (l+1)NB. \end{aligned}$$

Hence, Equation (4.18) must be valid for all $l \geq 1$.

Because every processor in the network can be reached from i by spanning fewer than N arcs, Equation (4.18) implies that

$$x_j(k') \geq \min_i\{x_i\} + (\gamma^{k'-k})^N[x_i - \min_i\{x_i\}] \tag{4.19}$$

for all $k' \geq k + N^2B$, $j \in p(i)$. Because we have made no assumptions to the contrary, Equation (4.19) is valid for any $i \in L$. Hence, we can replace x_i with $\max_i\{x_i\}$ and $j \in p(i)$ with $j \in L$ and Equation (4.19) becomes

$$x_j(k') \geq \min_i\{x_i\} + (\gamma^{k'-k})^N[\max_i\{x_i\} - \min_i\{x_i\}]$$

for all $k' \geq k + N^2B$, $j \in L$. It follows directly that

$$\min_i\{x_i(k')\} \geq \min_i\{x_i\} + (\gamma^{k'-k})^N[\max_i\{x_i\} - \min_i\{x_i\}] \tag{4.20}$$

for all $k' \geq k + N^2B$.

Choose $k' = k+N^2B$. For every $k \geq 0$, $\boldsymbol{x}_k \notin \mathcal{X}_b$, Equations (4.5) and (4.20) imply that

$$\begin{aligned} V(\boldsymbol{x}_k) - V(\boldsymbol{x}_{k+N^2B}) &= \min_i\{x_i(k+N^2B)\} - \min_i\{x_i\} \\ &\geq (\gamma^{N^2B})^N[\max_i\{x_i\} - \min_i\{x_i\}] \\ &\geq \gamma^{N^3B}\rho(\boldsymbol{x}_k, \mathcal{X}_b). \end{aligned} \tag{4.21}$$

The above equation satisfies the final condition of the Theorem 3.11. ■

Consider the following issues:

- The proof of Theorem 4.2 depends critically upon the fact that $\boldsymbol{E}_B$ requires that for every $i \in L$, the corresponding partial event, $e^{i,p(i)}$, occur at least once in every B events. Hence, it is clear that $\mathcal{X}_b$ is not exponentially stable in the large with respect to $\boldsymbol{E}_i$.

- In the proof, it is shown that

$$V(\boldsymbol{x}_k) - V(\boldsymbol{x}_{k+N^2B}) \geq \gamma^{N^3B}\rho(\boldsymbol{x}_k, \mathcal{X}_b) , \tag{4.22}$$

where

$$V(\boldsymbol{x}_k) = \max_i \left\{ \frac{1}{N} \sum_{j=1} N x_j - x_i \right\}.$$

The constant $\gamma^{N^3 B}$ from Equation (4.22) is directly related to the α from the exponential overbounding function $\zeta e^{-\alpha k} \rho(\boldsymbol{x}_0, \mathcal{X}_b)$. Thus, if speed of convergence is a design factor, then γ should be made as large as possible and N and B should be made as small as possible.

It is evident that Equation (4.22) is unnecessarily conservative. Equation (4.11), restated here,

$$x_j(k') \geq \min_i\{x_i\} + \gamma^{k'-k}[x_i - \min_i\{x_i\}] \text{ for all } k' \geq k + NB\ , \quad j \in p(i)\ ,$$

is also unnecessarily conservative. Actually, Equation (4.11) is valid for all $k' \geq k + RB$, where $R = \max_i\{|p(i)|\}$. Let S be the maximum number of arcs that must be spanned to reach any node $j \in L$ from any other node $i \in L$. N can be replaced by S in Equation (4.19), restated here,

$$x_j(k') \geq \min_i\{x_i\} + (\gamma^{k'-k})^N [x_i - \min_i\{x_i\}],$$

and Equation (4.22) becomes

$$V(\boldsymbol{x}_k) - V(\boldsymbol{x}_{k+RSB}) \geq \gamma^{RS^2 B} \rho(\boldsymbol{x}_k, \mathcal{X}_b).$$

Therefore, convergence can be accelerated by designing for RS^2 as small as possible.

Consider three common network topologies of N nodes. If N nodes are connected in a *line*, then $R = 2$ and $S = N - 1$. If N nodes are connected in a simple *ring*, then $R = 2$ and $S =$ int$(N/2)$ (int(x) is the integer portion of x). If N nodes are completely connected (each node is connected to every other node), then $R = N - 1$ and $S = 1$. In general, the ring network will converge more rapidly than the line network, and the completely connected network will converge more quickly than the ring network. Intuitively, this is what we would expect; convergence performance seems directly related to $|A|$.

- If we change our assumptions regarding the network topology to allow networks that are *strongly connected*, the above analysis may be simply amended to remain valid. We must replace N in Equation (4.19) with S, where S is defined above. Equation (4.22) must then be changed by replacing N^3 with $S^2 N$. If $S > N$, then the guaranteed rate of convergence for a strongly connected network with N nodes is slower than for a network with N nodes that satisfies our original network topology assumption. However, if the cost of internode connections is great, the sacrifice in convergence speed may be worthwhile. $|A|$ for a strongly connected network of N nodes has a minimum value of N, when the nodes are joined in a ring such that if $(i, j) \in A$ then $(j, i) \notin A$.

4.3 LOAD BALANCING PROBLEM GENERALIZATIONS

In this section we discuss generalizations of the load balancing problem previously outlined. First, we discuss less restrictive conditions on the amount of load that can be passed from node to node, coupled with a new specification of $\boldsymbol{E}_a$. Second, we discuss the idea of virtual load, a mechanism to account for the varied rates at which internetwork processors may process load. Finally, we consider the case in which the load in the network cannot be accurately modeled by a continuous variable (i.e., the discrete load case).

4.3.1 Generalized Load Passing Conditions

We will require that condition (i) on $e_k \in g(\boldsymbol{x}_k)$ remain unchanged. We change condition (ii), however, to allow that if $e^{i,p(i)}_{\alpha(i)} \in e_k$, then possibly after the passing of $\alpha(i)$, the load of node i can fall to the level of some node $j' \in p(i)$. This new condition (ii) is

$$(iia) \qquad 0 \leq \sum_{m \in p(i)} \alpha_m(i) \leq x_i - (x_{j'} + \alpha_{j'}(i)) \text{ for some } j' \in \{j : x_j < x_i , \ j \in p(i)\}$$

Condition (iii) is also changed because we no longer require that if $e^{i,p(i)}_{\alpha(i)} \in e_k$ then node i pass a nonnegligible amount of load to some least loaded neighbor j^*. The γ_{ij} are fixed *a priori* as before. The new condition (iii) is

$$(iiia) \qquad \alpha_{j'}(i) \geq \gamma_{ij'}(x_i - x'_j) \text{ for some } j' \in p(i) \text{ such that } x_{j'} < x_i.$$

Notice that we now require only that if $e^{i,p(i)}_{\alpha(i)} \in e_k$, then node i pass a predefined fraction of the load difference between nodes i and j' to node j' for *some* node $j' \in p(i)$.

We now define new sets of allowed event trajectories. We define an *elementary event*, $e^{ij}_{\alpha_j(i)}$ to represent the passing of load $\alpha_j(i)$ from processor i to processor j (note that $e^{i,p(i)}_{\alpha(i)} = \{e^{ij}_{\alpha_j(i)} : j \in p(i)\}$). We define an elementary event of type (i,j) to be any $e^{ij}_{\alpha_j(i)}$, and denote an elementary event of type (i,j) with e^{ij}.

(i) For $\boldsymbol{E}_I \subset \boldsymbol{E}_v$, every event trajectory $E \in \boldsymbol{E}_I$ must contain an infinite number of occurrences of elementary events of every type e^{ij} for all $(i,j) \in A$.

(ii) For $\boldsymbol{E}_{B'} \subset \boldsymbol{E}_v$, assume that there exists $B' > 0$, such that for every event trajectory $E \in \boldsymbol{E}_{B'}$, in every substring $e_{k'}, e_{k'+1}, e_{k'+2}, \ldots, e_{k'+(B'-1)}$ of E there is the occurrence of every type of elementary event (i.e. for every $i \in L$ elementary event $e^{ij} \in e_k$ for some k, $k' \leq k \leq k' + B' - 1$).

Theorem 4.3 (General Load Passing, No Delays, Continuous Load, Asymptotic Convergence): *For the load processor network system with conditions (iia) and (iiia) the invariant set $\mathcal{X}_b$ is asymptotically stable in the large with respect to $\boldsymbol{E}_I$.*

Proof: Using the same ρ and V as in Theorem 4.1, the proof for stability in the sense of Lyapunov with respect to $\boldsymbol{E}_I$ is the same as in the proof of Theorem 4.1. The proof of asymptotic stability in the large, however, must be slightly modified. In the original proof, we are guaranteed that the partial event $e^{i,p(i)}$, where $(i, j^{**}) \in A$ and $x_i > x_{j^{**}}$, must occur infinitely often. Given the above generalizations, we can simply state that the elementary event $e^{ij^{**}}$, where $x_i > x_{j^{**}}$, must occur infinitely often or until $\boldsymbol{x}_k \in \mathcal{X}_b$. Thus, we can say that the overall lightest load in the network must definitely increase an infinite number of times or until it is equal to the average network load. Hence, we have that for the generalized load system, $\mathcal{X}_b$ is asymptotically stable in the large with respect to $\boldsymbol{E}_I$. ∎

The new conditions allow for greater efficiency because it is no longer necessary for node i to examine all x_j with $j \in p(i)$ to find x_{j^*} before passing. In a network where $|p(i)|$ is large, this may prove to be quite a time-saving advantage.

Theorem 4.4 (General Load Passing, No Delays, Continuous Load, Exponential Stability): *For the load processor network system with conditions (iia) and (iiia) the invariant set $\mathcal{X}_b$ is exponentially stable in the large with respect to $\boldsymbol{E}_{B'}$.*

The proof is omitted, as it is very similar to the proof of Theorem 4.2.

4.3.2 Virtual Load

In practice, it is often the case that the load processors in the network may process the load at different rates. In this case, it is useful to scale the physical load of each processor by assigning constants $\beta_i > 0$, which are inversely proportional to the rate at which processor i can process load, for each $i \in L$. Hence, we define $\beta_i x_i$ as the *virtual* load of processor i, and it is the virtual load that we wish to balance among the network nodes. It is useful to balance the virtual load in a load processor network to ensure that nodes which process load faster have a larger portion of the available load.

With a few adjustments, the above analysis applies directly in the case of virtual load. First of all, because we are interested in balancing the virtual load, we should only allow node i to pass load to node j if the virtual load of node i is greater than the virtual load of node j. Accordingly, condition (i) on $e_k \in g(\boldsymbol{x}_k)$ must be changed to

$$(ib) \qquad \alpha_j(i) = 0 \text{ if } \beta_i x_i \le \beta_j x_j \text{ where } j \in p(i).$$

Secondly, we require that after node i passes load, its virtual load be at least as large as the possibly increased virtual load, due to $\alpha_{j^*}(i)$, of node j^*. This requirement can be expressed as

$$\beta_i \left(x_i - \sum_{j \in p(i)} \alpha_j(i) \right) \geq \beta_{j^*}(x_{j^*} + \alpha_{j^*}(i))$$

Direct manipulation of this equation leads to the extension of condition (iia)

$$(iib) \qquad 0 \leq \sum_{j \in p(i)} \alpha_j(i) \leq x_i - \frac{\beta_{j^*}}{\beta_i}(x_{j^*} + \alpha_{j^*}(i)) \text{ for all }$$
$$j^* \in \{j : \beta_j x_j \leq \beta_m x_m \text{ for all } m \in p(i)\}$$

We also must require that if node i is not virtual load balanced with all of its neighbors, then i must pass a nonnegligible portion of its load to at least one of its neighbors. We can express this condition as

$$\frac{1}{2}\{(\beta_i x_i - \beta_j x_j) - [\beta_i(x_i - \alpha_j(i)) - \beta_j(x_j + \alpha_j(i))]\} \geq \gamma_{ij}(\beta_i x_i - \beta_j x_j)$$

for some $j \in p(i)$. After some manipulation, we arrive at the virtual load version of condition $(iiia)$

$$(iiib) \qquad \alpha_j(i) \geq \frac{2\gamma_{ij}(\beta_i x_i - \beta_j x_j)}{\beta_i + \beta_j} \quad \text{for some } j \in p(i)$$

Notice that in the case of $\beta_i = 1$ for all $i \in L$, the conditions (ib), (iib), and $(iiib)$ properly reduce to conditions (i), (iia), and $(iiia)$.

Clearly,

$$\mathcal{X}_{bv} = \{\boldsymbol{x}_k \in \mathcal{X} : |\beta_i x_i - \beta_j x_j| = 0 \text{ for all } (i,j) \in A\}$$

is an invariant set which represents a perfectly balanced virtual load.

Theorem 4.5 (Continuous Virtual Load, No Delays, Exponential Stability): *For the virtual load processor network system with conditions (ib), (iib), and $(iiib)$ the invariant set $\mathcal{X}_{bv}$ is exponentially stable in the large with respect to $\boldsymbol{E}_{B'}$.*

If all references to x_i are replaced by references to $\beta_i x_i$ and the new conditions on $e_k \in g(\boldsymbol{x}_k)$ are observed, the proof is very similar to the proof of Theorem 4.2. In the virtual load balancing problem, it is of course necessary that node i not only have knowledge of x_j for all $j \in p(i)$, but also of β_j for all $j \in p(i)$. Just as new load can enter the load balanced system, perturbing the balance, the load processing capabilities of the load processors may change, perturbing the balance of the virtual load balanced system. Given that the $\beta = \{\beta_i : i \in L\}$ is updated to reflect the change in load processing capability (e.g. a change in the rate at which some node can process load), the system will recognize the imbalance and begin to rebalance from a new state $\boldsymbol{x}_0 \notin \mathcal{X}_{bv}$.

4.3.3 Discrete Load

Consider now that we have the same system as originally described, except that in this case, we may not assume that the load can be described with a continuous variable, as is the case in many practical systems. In fact, we assume that the load in the system is partitioned into blocks. The largest block in the network has size $M > 0$ and the smallest block in the network has size m, $M \geq m > 0$. In contrast to the perfect load balancing that is possible in the continuous load case, the best we can generally hope to do with only local information in the discrete load case is to balance each interprocessor connection to within M. Next, we define the model G for the discrete load case.

We utilize the same $\mathcal{X}$ and $\mathcal{E}$ as in the continuous load case. Below, we specify g and f_e for $e_k \in g(\boldsymbol{x}_k)$:

- Event $e_k \in g(\boldsymbol{x}_k)$ if both (a) and (b) below hold:

  (a) For all $e^{i,p(i)}_{\alpha(i)} \in e_k$, where $\alpha(i) = (\alpha_j(i) \ : \ j \in p(i))$ it is the case that:

  $$(i) \qquad \alpha_j(i) = 0, \ \text{ if } x_i - x_j \leq M \ \text{ where } j \in p(i),$$

  $$(ii) \qquad x_i - \sum_{m \in p(i)} \alpha_m(i) > \min_j\{x_j \ : \ j \in p(i)\}$$

  $$(iii) \qquad \text{If } \ \alpha_j(i) > 0, \ \text{ for some } j \in p(i) \ , \ \text{ then}$$
  $$\alpha_{j^*}(i) > 0, \ \text{ for some } j^* \in \{j : \ x_j \leq x_m \ \text{ for all } m \in p(i)\}.$$

  Condition (i) prevents load from being passed by node i to node j if nodes i and j are balanced within M. Condition (ii) implies that after the load $\alpha(i)$ has been passed, the remaining load of node i must be larger than the load at time k of some neighbor of i. Condition (iii) implies that if node i is not load balanced to within M with all of its neighbors, then i must pass some load to some least loaded neighbor j^*.

  (b) If $e^{i,p(i)}_{\alpha(i)} \in e_k$, where $\alpha(i) = (\alpha_j(i) \ : \ j \in p(i))$, then $e^{i,p(i)}_{\delta(i)} \notin e_k$ where $\delta(i) = \{\delta_j(i) \ : \ j \in p(i)\}$ if $\alpha_j(i) \neq \delta_j(i)$ for some $j \in p(i)$. Hence, in each valid event e_k, there must be a consistent definition of the load to be passed from any node i to any other node j, $\alpha_j(i)$.

- If $e_k \in g(\boldsymbol{x}_k)$ and $e^{i,p(i)}_{\alpha(i)} \in e_k$, then $f_{e_k}(\boldsymbol{x}_k) = \boldsymbol{x}_{k+1}$, where

  $$x'_i = x_i - \sum_{\{j: \, j \in p(i)\}} \alpha_j(i) + \sum_{\{j: \, i \in p(j) \ , \ e^{j,p(j)}_{\alpha(j)} \in e_k\}} \alpha_i(j).$$

  The load of node i at time $k+1$, x'_i, is the load of node i at time k minus the total load passed by node i at time k plus the total load received by node i at time k.

Let $\boldsymbol{E}_v = \boldsymbol{E}$ be the set of valid event trajectories. Define $\boldsymbol{E}_i, \boldsymbol{E}_B \subset \boldsymbol{E}_v$ as in the continuous load case.

Clearly,

$$\mathcal{X}_{bd} = \{\boldsymbol{x_k} \in \mathcal{X} : |x_i - x_j| \leq M \text{ for all } (i,j) \in A\}$$

is an invariant set that represents a balanced load in the sense described above. Notice that the only $e_k \in g(\boldsymbol{x}_k)$, where $\boldsymbol{x}_k \in \mathcal{X}_{bd}$, are ones such that all $e_{\alpha(i)}^{i,p(i)} \in e_k$ have $\alpha(i) = (0, 0, \ldots, 0)$.

Asymptotic Convergence to an M Balanced State Once again, we employ a Lyapunov stability theoretic approach. Let $\bar{\boldsymbol{x}} = [\bar{x}_1, \ldots, \bar{x}_N]$. Choose

$$\rho(\boldsymbol{x}_k, \mathcal{X}_{bd}) = \inf\{\max\{|x_i - \bar{x}_i| : i \in L\} : \bar{\boldsymbol{x}} \in \mathcal{X}_{bd}\}. \tag{4.23}$$

Theorem 4.6 (Discrete Load, No Delays, Asymptotic Convergence): *For the discrete load processor network system, the invariant set $\mathcal{X}_{bd}$ is asymptotically stable in the large with respect to $\boldsymbol{E}_i$.*

Proof: Choose

$$V(\boldsymbol{x}_k) = \begin{cases} \frac{1}{N}\sum_{i=1}^{N} x_i - \min_i\{x_i\} & , \quad \boldsymbol{x}_k \notin \mathcal{X}_{bd} \\ 0 & , \quad \boldsymbol{x}_k \in \mathcal{X}_{bd} \end{cases} \tag{4.24}$$

Notice that for $\boldsymbol{x}_k \notin \mathcal{X}_{bd}$, there must be two nodes i and j, $(i,j) \in A$, such that $x_i - x_j > M$. Because nodes i and j, $(i,j) \in A$, of any state $\bar{\boldsymbol{x}} \in \mathcal{X}_{bd}$ must be such that $\bar{x}_i - \bar{x}_j \leq M$, it is clear from Equation (4.23) that

$$\begin{aligned} \rho(\boldsymbol{x}_k, \mathcal{X}_{bd}) &\geq \frac{1}{2}\max\{x_i - x_j - M : (i,j) \in A\} \\ 2\rho(\boldsymbol{x}_k, \mathcal{X}_{bd}) &\geq \max\{x_i - x_j - M : (i,j) \in A\} \triangleq \psi_1(\boldsymbol{x}_k). \end{aligned} \tag{4.25}$$

According to Equation (4.24), because $\max_i\{x_i\} \geq \frac{1}{N}\sum_{i=1}^{N} x_i$, $V(\boldsymbol{x}_k) \leq \max_i\{x_i\} - \min_i\{x_i\}$. Because there exists a network link between any two nodes that consists of fewer than N interprocessor connections, it must be true that $\max_i\{x_i\} - \min_i\{x_i\} \leq N \max\{x_i - x_j : (i,j) \in A\}$. Hence,

$$\begin{aligned} V(\boldsymbol{x}_k) &\leq N \max\{x_i - x_j : (i,j) \in A\} \\ \frac{1}{N} V(\boldsymbol{x}_k) &\leq \max\{x_i - x_j : (i,j) \in A\} \triangleq \psi_2(\boldsymbol{x}_k). \end{aligned} \tag{4.26}$$

Finally, notice that according to Equations (4.23) and (4.24),

$$V(\boldsymbol{x}_k) = \rho(\boldsymbol{x}_k, \mathcal{X}_{bd}) = 0 , \quad \boldsymbol{x}_k \in \mathcal{X}_{bd}. \tag{4.27}$$

We will find a constant $\eta \in (0, \infty)$ such that $\eta\rho(\boldsymbol{x}_k, \mathcal{X}_{bd}) \geq V(\boldsymbol{x}_k)$ for all $\boldsymbol{x}_k \in \mathcal{X}$. From Equation (4.27), we see that for all $\boldsymbol{x}_k \in \mathcal{X}_b$, any value of η

will suffice. Thus, we need only be concerned with $\boldsymbol{x}_k \notin \mathcal{X}_{bd}$. Accordingly, we will find a constant $\phi \in (0, \infty)$ such that $\phi\psi_1(\boldsymbol{x}_k) \geq \psi_2(\boldsymbol{x}_k)$ for $\boldsymbol{x}_k \notin \mathcal{X}_{bd}$. Notice from Equations (4.25) and (4.26) that

$$\psi_1(\boldsymbol{x}_k) + M \;=\; \psi_2(\boldsymbol{x}_k)$$

and $\psi_1(\boldsymbol{x}_k) \;=\; \epsilon$, $\epsilon > 0$, implies that $\psi_2(\boldsymbol{x}_k) \;=\; \epsilon + M$. From this, it is clear that $\phi \in [1, \infty)$. For very large values of ϵ, a value of ϕ close to unity will satisfy our requirement, but as ϵ approaches zero, the necessary value of ϕ approaches infinity. However, because the network contains a finite number of blocks, each of finite size, there must be some constant, ϵ_0, $M \geq \epsilon_0 > 0$, such that for $\boldsymbol{x}_k \notin \mathcal{X}_{bd}$, $\psi_1(\boldsymbol{x}_k) \geq \epsilon_0$. Thus, if we choose $\phi = 2M/\epsilon_0$, it is clear that if $\psi_1(\boldsymbol{x}_k) \geq M$, then

$$\phi\psi_1(\boldsymbol{x}_k) \;=\; \frac{2M}{\epsilon_0}\psi_1(\boldsymbol{x}_k) \;\geq\; 2\psi_1(\boldsymbol{x}_k) \;\geq\; \psi_1(\boldsymbol{x}_k) + M \;=\; \psi_2(\boldsymbol{x}_k),$$

and if $\psi_1(\boldsymbol{x}_k) < M$, then

$$\phi\psi_1(\boldsymbol{x}_k) \;=\; \frac{2M}{\epsilon_0}\psi_1(\boldsymbol{x}_k) \;\geq\; 2M \;\geq\; \psi_1(\boldsymbol{x}_k) + M \;=\; \psi_2(\boldsymbol{x}_k).$$

It follows that, for all $\psi_1(\boldsymbol{x}_k) \geq \epsilon_0$,

$$\phi\psi_1(\boldsymbol{x}_k) \geq \psi_2(\boldsymbol{x}_k). \tag{4.28}$$

From Equations (4.25), (4.26), (4.27), and (4.28), we see that

$$\begin{aligned} 2\phi\rho(\boldsymbol{x}_k, \mathcal{X}_{bd}) \;&\geq\; \phi\psi_1(\boldsymbol{x}_k) \geq \psi_2(\boldsymbol{x}_k) \geq \frac{1}{N}V(\boldsymbol{x}_k) \\ 2N\phi\rho(\boldsymbol{x}_k, \mathcal{X}_{bd}) \;&\geq\; V(\boldsymbol{x}_k) \;\text{ for all }\; \boldsymbol{x}_k \in \mathcal{X} \end{aligned} \tag{4.29}$$

so that condition (*ii*) of the Theorem 3.8 is satisfied.

Notice from Equation (4.23) that for $\boldsymbol{x}_k \notin \mathcal{X}_{bd}$,

$$\rho(\boldsymbol{x}_k, \mathcal{X}_{bd}) \leq \max_i\{x_i\} - \min_i\{x_i\}. \tag{4.30}$$

Because

$$\frac{1}{N}\sum_{i=1}^{N} x_i \geq \frac{1}{N}[\max_i\{x_i\} + (N-1)\min_i\{x_i\}],$$

we see from Equation (4.24) that

$$\begin{aligned} V(\boldsymbol{x}_k) \;&\geq\; \frac{1}{N}[\max_i\{x_i\} + (N-1)\min_i\{x_i\}] - \min_i\{x_i\} \\ &\geq\; \frac{1}{N}[\max_i\{x_i\} - \min_i\{x_i\}] \\ NV(\boldsymbol{x}_k) \;&\geq\; \max_i\{x_i\} - \min_i\{x_i\}. \end{aligned} \tag{4.31}$$

Thus, from Equations (4.27), (4.30) and (4.31), we conclude that

$$NV(\boldsymbol{x}_k) \geq \rho(\boldsymbol{x}_k, \mathcal{X}_{bd}) \text{ for all } \boldsymbol{x}_k \in \mathcal{X}, \tag{4.32}$$

so that condition (i) of the Theorem 3.8 is satisfied. Hence, condition (i) of Theorem 3.1 is satisfied.

Condition (ii) of the Theorem 3.1 is satisfied in exactly the same way as in the proof of Theorem 4.1 so that $\mathcal{X}_{bd}$ is stable in the sense of Lyapunov with respect to $\boldsymbol{E}_i$.

In order to show that $\mathcal{X}_{bd}$ is asymptotically stable in the large with respect to $\boldsymbol{E}_i$, we must show that for all $\boldsymbol{x}_0 \notin \mathcal{X}_{bd}$ and all E_k, such that $E_k E \in \boldsymbol{E}_i(\boldsymbol{x}_0)$,

$$V(X(\boldsymbol{x}_0, E_k, k)) \longrightarrow 0 \quad \text{as} \quad k \longrightarrow \infty. \tag{4.33}$$

If $\boldsymbol{x}_k \notin \mathcal{X}_{bd}$, then there must be some lightest loaded node j^{**} (there may be more than one such node) and some other node i such that $(i, j^{**}) \in A$ and $x_i > x_{j^{**}}$. Because of the restrictions imposed by $\boldsymbol{E}_i$, we know that all the partial events are guaranteed to occur infinitely often. According to condition $(a)(iii)$ on $e_k \in g(\boldsymbol{x}_k)$, each time partial event $e^{i,p(i)}$ occurs, $x_{j^{**}}$ is guaranteed to increase by m so that $x'_{j^{**}} \geq x_{j^{**}} + m$, and according to condition $(a)(ii)$ on $e_k \in g(\boldsymbol{x}_k)$, x'_i is guaranteed to be greater than $x_{j^{**}}$. In fact, because the system is composed of a finite number of blocks, each of finite size, we know that there is some constant $\delta > 0$ such that $x'_i \geq x_{j^{**}} + \delta$. Thus, regardless of how many lightest loaded nodes there are, it is inevitable that eventually the overall lightest load in the network must increase. Hence, for every $k \geq 0$, there exists $k' > k$ such that $V(\boldsymbol{x}_{k'}) > V(\boldsymbol{x}_{k'+1})$ as long as $\boldsymbol{x}_{k'} \notin \mathcal{X}_{bd}$ so that Equation (4.33) holds and $\mathcal{X}_{bd}$ is asymptotically stable in the large with respect to $\boldsymbol{E}_i$. ■

Exponential Convergence to an M Balanced State We employ Theorem 3.11 to prove that $\rho(\boldsymbol{x}_k, \mathcal{X}_{bd})$ is be bounded from above by an exponential

$$\zeta e^{-\alpha k} \rho(\boldsymbol{x}_0, \mathcal{X}_{bd})$$

for some $\alpha > 0$ and $\zeta > 0$.

Theorem 4.7 (Discrete Load, No Delays, Exponential Stability): *For the discrete load processor network system described above, the invariant set $\mathcal{X}_{bd}$ is exponentially stable in the large with respect to $\boldsymbol{E}_B$.*

Proof: The first condition of Theorem 3.11 is shown to hold in the proof of Theorem 4.6. We now show that the final condition of Theorem 3.11 holds.

We define a constant δ on which the proof will depend. For a given discrete load network, there is a constant $\delta_1 > 0$, such that if $e^{i,p(i)}_{\alpha(i)} \in e_k$ and $\alpha_j(i) > 0$ for some $j \in p(i)$, then $x'_i \geq x_{j^*} + \delta_1$, where $j^* \in p(i)$ and $x_{j^*} \leq x_j$ for all $j \in p(i)$. For the same discrete load network, there is also a constant $\delta_2 > 0$, such that if $(i, j) \in A$ and $x_i \neq x_j$, then $|x_i - x_j| \geq \delta_2$. Let $\delta = \min\{\delta_1, \delta_2, m\}$.

For $\boldsymbol{x}_k \notin \mathcal{X}_{bd}$, there is $L^*(k) \subset L$, such that $L^*(k) = \{i : x_i \leq x_j, j \in L\}$. Because there must be at least one node in the network that is more heavily loaded than the rest of the nodes, we know that $|L^*(k)| \leq N - 1$.

Fix a time $k \geq 0$. There must be some $i \notin L^*(k)$ and some $j \in L^*(k)$ such that $(i,j) \in A$. According to the restrictions imposed by $\boldsymbol{E}_B$, there is some time k_1, $k \leq k_1 < k + B$ such that $e_{\alpha(i)}^{i,p(i)} \in e_{k_1}$. Conditions $(a)(ii)$ and $(a)(iii)$ on $e_k \in g(\boldsymbol{x}_k)$, along with the definition of δ, imply that either (a) $|L^*(k+1)| \leq |L^*(k)| - 1$ and $x_q' = x_j$, for all $q \in L^*(k+1)$ and all $j \in L^*(k)$ or (b) $x_q' \geq x_j + \delta$ for all $q \in L^*(k+1)$ and all $j \in L^*(k)$. In other words, either the number of least loaded nodes decreases by at least one or the smallest load increases by at least δ. Thus, because $|L^*(k)| \leq N - 1$, we can conclude that for $\boldsymbol{x}_k \notin \mathcal{X}_{bd}$, $V(\boldsymbol{x}_k) - V(\boldsymbol{x}_{k+NB}) \geq \delta$. From Equation (4.23), it is clear that $\sum_{i=1}^N x_i > \rho(\boldsymbol{x}_k, \mathcal{X}_{bd})$. It is also clear that there is some $\zeta > 0$, such that

$$\zeta\delta > \sum_{i=1}^{N} x_i > \rho(\boldsymbol{x}_k, \mathcal{X}_{bd}).$$

Therefore, it follows that $V(\boldsymbol{x}_k) - V(\boldsymbol{x}_{k+NB}) > \frac{1}{\zeta}\rho(\boldsymbol{x}_k, \mathcal{X}_{bd})$, which satisfies the final condition of Theorem 3.11. ■

Notice that in the discrete load case, the rate of exponential convergence depends on N, B and ζ. As in the continuous load case, the smaller we make B, the faster we are guaranteed to converge. Unlike the continuous load case in which the guaranteed rate of convergence depends on tangible system constants R and S, in this case we have the peculiar dependence on ζ. It less clear how to design for a small ζ than it is to design for a small R or S. If all the load blocks in the network have size M, then $\delta = M$ and

$$\zeta = \frac{\sum_{i=1}^N x_i}{M}.$$

However, if the load blocks are of various sizes then ζ must be calculated from a worst-case analysis.

It can be shown if we change condition $(a)(iii)$ to "if $x_i - x_j > M$ for some $(i,j) \in A$, then $\alpha_j(i) > 0$ for some $(i,j) \in A$", thereby alleviating the nodes from scanning all of their neighbors to locate one of the least loaded, that $\mathcal{X}_{bd}$ is asymptotically stable in the large with respect to $\boldsymbol{E}_i$ and exponentially stable in the large with respect to $\boldsymbol{E}_B$. Of course, the guaranteed rate of convergence will suffer under this less strict load passing condition.

4.4 THE LOAD BALANCING PROBLEM WITH DELAYS

We now modify the model of the system to allow for delays in load transport and sensing. In this extended analysis, we no longer require that the *real time* between events e_k and e_{k+1} be greater than the greatest system transportation time plus the greatest system sensing time. In this sense, we allow a reduction of the degree of synchronicity forced upon the system. What we now require is that there exist $B > 0$ such that load passed at time k is received by time $k + B - 1$ and that for all $(i, j) \in A$ load which arrives at node j at time k' will be sensed by node i by time $k' + B - 1$.

First, we specify the model G. Let $\mathcal{X} = \Re^{(2N+|A|)\times B}$ be the set of states. Every $\boldsymbol{x}_k \in \mathcal{X}$ is composed of three "substates." Let $\boldsymbol{x}_{n0} \in \Re^{N\times B}$ represent the loads of the N network nodes at times $k, k-1, \ldots, k-B+2, k-B+1$. The first column represents the loads of the nodes at time k, the second column represents the loads of the nodes at time $k-1$, and so on. Let $\boldsymbol{x}_{n1} \in \Re^{N\times B}$ represent the loads of the N network nodes at times $k-B, k-B-1, \ldots, k-2B+2, k-2B+1$. The first column represents the loads of the nodes at time $k-B$, the second column represents the loads of the nodes at time $k-B-1$, and so on. Let $\boldsymbol{x}_t \in \Re^{|A|\times B}$ represent all of the $|A|$ loads in transit between the N network nodes at times $k, k-1, \ldots, k-B+2, k-B+1$. The first column represents the loads in transit at time k, the second column represents the loads in transit at time $k-1$, and so on. Pictorially, the state $\boldsymbol{x}_k \in \mathcal{X}$ may be represented as

$$\boldsymbol{x}_k = \begin{bmatrix} \boldsymbol{x}_{n0} \\ \boldsymbol{x}_{n1} \\ \boldsymbol{x}_t \end{bmatrix} = \begin{bmatrix} \boldsymbol{x}_n \\ \boldsymbol{x}_t \end{bmatrix} \quad \text{where} \quad \boldsymbol{x}_n = \begin{bmatrix} \boldsymbol{x}_{n0} \\ \boldsymbol{x}_{n1} \end{bmatrix}.$$

We also define

$$\boldsymbol{x}_s = \begin{bmatrix} \boldsymbol{x}_{n0} \\ \boldsymbol{x}_t \end{bmatrix},$$

so that the sum of the elements of any column of $\boldsymbol{x}_s$ is equal to the total network load. Let $\boldsymbol{x}_{n0}(k')$, $\boldsymbol{x}_{n1}(k')$, $\boldsymbol{x}_n(k')$, $\boldsymbol{x}_t(k')$, and $\boldsymbol{x}_s(k')$ be defined in the same manner as $\boldsymbol{x}_{n0}$, $\boldsymbol{x}_{n1}$, $\boldsymbol{x}_n$, $\boldsymbol{x}_t$, and $\boldsymbol{x}_s$, with the exception that the state from which they derive is $\boldsymbol{x}_{k'}$ instead of $\boldsymbol{x}_k$.

Let x_i denote the load of node $i \in L$ at time k, let x_i' denote the load of node $i \in L$ at time $k+1$, and let $x_i(k')$ denote the load of node $i \in L$ at time k'. Clearly, x_i is element i of the first column of $\boldsymbol{x}_k$, x_i' is element i of the first column of $\boldsymbol{x}_{k+1}$, and $x_i(k')$ is element i of the first column of $\boldsymbol{x}_{k'}$. Let $x_{i\to j}$ denote the load in transit from node i to node j at time k, and let $x_{i\to j}'$ denote the load in transit from node i to node j at time $k+1$, $(i, j) \in A$. Clearly, $x_{i\to j}$ is one of the last $|A|$ elements of the first column of $\boldsymbol{x}_k$, and $x_{i\to j}'$ is one of the last $|A|$ elements of the first column of $\boldsymbol{x}_{k+1}$. Let x_j^i be the perception by node i of the load of node j at time k, and let $x_j^i(k')$ be the perception by node i of the load of node j at time k'. Because of the restriction on the delay in sensing, x_j^i must be *any one* of the elements of row j of $\boldsymbol{x}_k$ (i.e., row j of $\boldsymbol{x}_{n0}$), and $x_j^i(k')$ must be *any one* of the elements of row j of $\boldsymbol{x}_{k'}$.

Let $e_{\alpha(i)}^{i\to p(i)}$ represent that node $i \in L$ passes load to its neighbors $m \in p(i)$, where $p(i) = \{j : \exists\ (i, j) \in A\}$. Let $\alpha(i) = (\alpha_j(i), \alpha_{j'}(i), \ldots, \alpha_{j''}(i))$, such that $j < j' < \cdots < j''$ and $j, j', \ldots, j'' \in p(i)$ and $\alpha_j \geq 0$ for all $j \in p(i)$; the size of the list is $|p(i)|$. For convenience, we will denote this list by $\alpha(i) = (\alpha_j(i) : j \in p(i))$. $\alpha_m(i)$ denotes the amount of load passed from $i \in L$ to $m \in p(i)$. Let $\{e^{i \to p(i)}_{\alpha(i)}\}$ denote the set of all possible such load passes. Let $e^{j \leftarrow i}_{\beta}$ represent that node $j \in L$ receives $\beta \geq 0$ load from node i. Let $\{e^{j \leftarrow i}_{\beta}\}$ denote the set of all possible such load receptions. Let the set of events be described by

$$\mathcal{E} = \left\{\mathcal{P}\left(\{e^{i \to p(i)}_{\alpha(i)}\}\right) \bigcup \mathcal{P}\left(\{e^{j \leftarrow i}_{\beta}\}\right)\right\} - \{\phi\}.$$

As before, each event $e_k \in \mathcal{E}$ is defined as a set. Elements of e_k may represent either the passing of load by node $i \in L$ to its neighboring nodes in the network or the reception of load by node $i \in L$. Once again, let the γ_{ij} for $(i, j) \in A$ be defined *a priori.*

Below, we specify g and f_e for $e_k \in g(\boldsymbol{x}_k)$:

- Event $e_k \in g(\boldsymbol{x}_k)$ if (a), (b), and (c) below hold:

  (a) For all $e^{i \to p(i)}_{\alpha(i)} \in e_k$, where $\alpha(i) = (\alpha_j(i) : j \in p(i))$ it is the case that:

  (i) $\alpha_j(i) = 0$ if $x_i \leq x^i_j$ where $j \in p(i)$,

  (ii) $0 \leq \sum_{m \in p(i)} \alpha_m(i) \leq x_i - (x^i_j + \alpha_j(i))$ for all $j \in p(i)$, such that $x_i \geq x^i_j$ and

  (iii) $\alpha_{j^*}(i) \geq \gamma_{ij^*}(x_i - x^i_{j^*})$ for some $j^* \in \{j : x^i_j \leq x^i_m$ for all $m \in p(i)\}$.

  Condition (i) prevents load from being passed by node i to node j if node i is less heavily loaded than its perception of node j. Condition (ii) directly implies that $x_i - \sum_{m \in p(i)} \alpha_m(i) \geq x^i_j + \alpha_j(i)$. Thus, after the load $\alpha(i)$ has been passed, the remaining load of node i must be at least as large as $x^i_j + \alpha_j(i)$ for every node $j \in p(i)$ that was less heavily loaded than node i to begin with. Condition (iii) implies that if node i does not perceive itself as being load balanced with all of its neighbors, then i must pass a nonnegligible portion of its load to some neighbor perceived to be least loaded, j^*.

  (b) For all $e^{j \leftarrow i}_{\beta} \in e_k$ it is the case that $0 \leq \beta \leq x_{i \to j}$.

  (c) If $e^{i \to p(i)}_{\alpha(i)} \in e_k$, where $\alpha(i) = (\alpha_j(i) : j \in p(i))$, then $e^{i \to p(i)}_{\delta(i)} \notin e_k$, where $\delta(i) = (\delta_j(i) : j \in p(i))$ if $\alpha_j(i) \neq \delta_j(i)$ for all $j \in p(i)$. Hence, in each valid event e_k, there must be a consistent definition of the load, $\alpha_j(i)$, to be passed from any node i to any other node j.

(d) If $e_\beta^{j\leftarrow i} \in e_k$, then $e_\delta^{j\leftarrow i} \notin e_k$ if $\beta \neq \delta$. Hence, in each valid event e_k, there must be a consistent definition of the load, β, received by any node j from any other node i, $(i,j) \in A$.

- If $e_k \in g(\boldsymbol{x}_k)$, then $f_{e_k}(\boldsymbol{x}_k) = \boldsymbol{x}_{k+1}$, where

$$
\begin{aligned}
x'_i &= x_i - \sum_{\{j:\, j \in p(i),\, e_{\alpha(i)}^{i\rightarrow p(i)}\}} \alpha_j(i) + \sum_{\{m:\, e_\beta^{i\leftarrow m} \in e_k\}} \beta, \\
x'_{i\rightarrow j} &= x_{i\rightarrow j} + \sum_{\{j:\, j \in p(i),\, e_{\alpha(i)}^{i\rightarrow p(i)} \in e_k\}} \alpha_j(i) - \sum_{\{j:\, e_\beta^{i\leftarrow j} \in e_k\}} \beta.
\end{aligned}
$$

The load of node i at time $k+1$ is the load of node i at time k minus the total load passed by node i at time k plus the total load received by node i at time k. The load in transit from node i to any one of its neighbors, $j \in p(i)$, at time $k+1$ is the load in transit from node i to node j at time k, plus the passed load, minus the received load.

Let $\boldsymbol{E}_v = \boldsymbol{E}$ be the set of valid event trajectories. We must further specify the set of allowed event trajectories, $\boldsymbol{E}_a \subset \boldsymbol{E}_v$. We define a *partial event* of type "$i \rightarrow$" to represent the passing of $\alpha(i)$ amount of load from $i \in L$ to its neighbors $p(i)$. A partial event of type $i \rightarrow$ will be denoted by $e^{i\rightarrow p(i)}$ and the occurrence of $e^{i\rightarrow p(i)}$ indicates that $i \in L$ attempts to further balance its load with its neighbors. We define a partial event of type "$j \leftarrow$" to represent the receiving of β amount of load by $j \in L$ from one of its neighbors in $p(j)$. Event e_k is composed of a set of partial events. For $\boldsymbol{E}_B \subset \boldsymbol{E}_v$, the following two conditions must hold for every $E \in \boldsymbol{E}_B$:

- There exists $B > 0$, such that in every substring

$$e_{k'}, e_{k'+1}, e_{k'+2}, \ldots, e_{k'+(B-1)}$$

there is the occurrence of partial event $e^{i\rightarrow p(i)}$ for all $i \in L$ (i.e., for every $i \in L$ partial event $e^{i\rightarrow p(i)} \in e_k$ for some k, $k' \leq k \leq k' + B - 1$).

- For every i and k' such that $e_{\alpha(i)}^{i\rightarrow p(i)} \in e_{k'}$, there is $k' \leq k < k' + B$, such that $e_{\alpha_j(i)}^{j\leftarrow i} \in e_k$. This restriction mandates that load passed at time k' must be received intact by time $k' + B - 1$.

We want to define an invariant set, such that any state $\boldsymbol{x}_k$ that is in the invariant set exhibits the following properties:

(*i*) The load in the nodes is perfectly balanced at time k;

(*ii*) There is no load in transit at time k; and

(*iii*) At time k, every node has an accurate perception of the load of its neighbors.

Let $L' = \{1, 2, \ldots, 2N\}$, $G = \{1, 2, \ldots, B\}$, and $H = \{1, 2, \ldots, |A|\}$. If $\boldsymbol{y}$ is a matrix, let $(\boldsymbol{y})_{pq}$ denote the element in row p and column q of $\boldsymbol{y}$. Choose

$$\begin{aligned} \mathcal{X}_b = \quad & \{\boldsymbol{x}_k \in \mathcal{X} : (\boldsymbol{x}_n)_{ij} = (\boldsymbol{x}_n)_{pq} \text{ for all } i, p \in L' \text{ and } j, q \in G; \\ & (\boldsymbol{x}_t)_{ij} = 0 \text{ for all } i \in H \text{ and } j \in G\}. \end{aligned} \tag{4.34}$$

Consider any $\boldsymbol{x}_k \in \mathcal{X}_b$. Because all elements of $\boldsymbol{x}_n$ are equal, the load in the nodes is perfectly balanced at time k. Because all elements of $\boldsymbol{x}_t$ are zero, there can be no load in transit at time k. Because the load at all nodes has been fixed since time $k - 2B + 1$, we are guaranteed that each node has an accurate perception of all of its neighbors at time k. Hence, $\mathcal{X}_b$ is an invariant set whose element states exhibit the required properties. Notice that the only $e_k \in g(\boldsymbol{x}_k)$, where $\boldsymbol{x}_k \in \mathcal{X}_b$, are ones such that all $e_{\alpha(i)}^{i \to p(i)} \in e_k$ have $\alpha(i) = (0, 0, \ldots, 0)$ and all $e_\beta^{j \leftarrow i} \in e_k$ have $\beta = 0$.

4.4.1 Asymptotic Convergence to a Balanced State

To study the ability of the system to automatically redistribute load to achieve balancing, we again employ a Lyapunov stability theoretic approach. Let $T = \{1, 2, \ldots, (2N + |A|)\}$. Choose

$$\rho(\boldsymbol{x}_k, \mathcal{X}_b) = \inf\{\max\{|(\boldsymbol{x}_k)_{ij} - (\bar{\boldsymbol{x}})_{ij}| : \text{ for all } i \in T\ ,\ \ j \in G\} : \bar{\boldsymbol{x}} \in \mathcal{X}_b\}. \tag{4.35}$$

Theorem 4.8 (Continuous Load, Delays, Asymptotic Convergence): *For the load processor network with delays as described above, the invariant set $\mathcal{X}_b$ is asymptotically stable in the large with respect to $\boldsymbol{E}_B$.*

Proof: For convenience, we define some mathematical notation. If $\boldsymbol{y}$ is a matrix, then $\min\{\boldsymbol{y}\}$ is equal to the minimum of all of the elements of $\boldsymbol{y}$, $\max\{\boldsymbol{y}\}$ is equal to the maximum of all of the elements of $\boldsymbol{y}$, and $\sum \boldsymbol{y}$ is equal to the sum of all of the elements of $\boldsymbol{y}$. Further, let $(\boldsymbol{y})^i$ be column i of $\boldsymbol{y}$.

Choose

$$V(\boldsymbol{x}_k) = \frac{1}{NB} \sum \boldsymbol{x}_s - \min\{\boldsymbol{x}_n\}. \tag{4.36}$$

Notice that $V(\boldsymbol{x}_k)$ is the average load (total network load divided by N) minus the minimum load, taken over times $k - 2B + 1, \ldots, k - 1, k$, at any node $i \in L$.

We first demonstrate that condition (*ii*) of Theorem 3.8 is satisfied by our choice of $\rho(\boldsymbol{x}_k, \mathcal{X}_b)$ and $V(\boldsymbol{x}_k)$.

It is clear from Equations (4.34) and (4.35) that

$$\rho(\boldsymbol{x}_k, \mathcal{X}_b) \quad \geq \quad \max\left\{\frac{1}{2}(\max\{\boldsymbol{x}_n\} - \min\{\boldsymbol{x}_n\})\ ,\ \max\{\boldsymbol{x}_t\}\right\}$$

$$\begin{aligned}\rho(\boldsymbol{x}_k, \mathcal{X}_b) &\geq \max\left\{\frac{1}{2}(\max\{\boldsymbol{x}_n\} - \min\{\boldsymbol{x}_n\})\ ,\ \frac{1}{2}\max\{\boldsymbol{x}_t\}\right\} \\ &\geq \frac{1}{2}\max\{(\max\{\boldsymbol{x}_n\} - \min\{\boldsymbol{x}_n\})\ ,\ \max\{\boldsymbol{x}_t\}\}. \qquad (4.37)\end{aligned}$$

It is also clear that

$$\max\{\boldsymbol{x}_k\} = \max\{\max\{\boldsymbol{x}_n\}\ ,\ \max\{\boldsymbol{x}_t\}\}.$$

We must consider two cases. If $\max\{\boldsymbol{x}_n\} - \min\{\boldsymbol{x}_n\} \geq \max\{\boldsymbol{x}_t\}$, then

$$\max\{\boldsymbol{x}_n\} \geq \max\{\boldsymbol{x}_t\} \quad \text{and} \quad \max\{\boldsymbol{x}_n\} = \max\{\boldsymbol{x}_k\}.$$

It follows, then, from Equation (4.37) that

$$2\rho(\boldsymbol{x}_k, \mathcal{X}_b) \geq \max\{\boldsymbol{x}_k\} - \min\{\boldsymbol{x}_n\}.$$

On the other hand, if $\max\{\boldsymbol{x}_t\} \geq \max\{\boldsymbol{x}_n\} - \min\{\boldsymbol{x}_n\}$, then because we know that

$$\max\{\boldsymbol{x}_t\} \geq \max\{\boldsymbol{x}_t\} - \min\{\boldsymbol{x}_n\}$$

and

$$\max\{\boldsymbol{x}_k\} = \max\{\max\{\boldsymbol{x}_t\}, \max\{\boldsymbol{x}_n\}\},$$

it must be the case that

$$\max\{\boldsymbol{x}_t\} \geq \max\{\boldsymbol{x}_k\} - \min\{\boldsymbol{x}_n\}.$$

Once again, Equation (4.37) implies that

$$2\rho(\boldsymbol{x}_k, \mathcal{X}_b) \geq \max\{\boldsymbol{x}_k\} - \min\{\boldsymbol{x}_n\}.$$

Thus, we can conclude that

$$2\rho(\boldsymbol{x}_k, \mathcal{X}_b) \geq \max\{\boldsymbol{x}_k\} - \min\{\boldsymbol{x}_n\} \qquad (4.38)$$

for all $\boldsymbol{x}_k \notin \mathcal{X}_b$.

If $\max\{\boldsymbol{x}_k\} \geq (1/NB)\sum \boldsymbol{x}_s$, then Equations (4.36) and (4.38) imply that

$$V(\boldsymbol{x}_k) \leq \max\{\boldsymbol{x}_k\} - \min\{\boldsymbol{x}_n\} \leq 2\rho(\boldsymbol{x}_k, \mathcal{X}_b). \qquad (4.39)$$

However, it is possible that some load is in transit and that the load at the nodes is distributed such that

$$\max\{\boldsymbol{x}_k\} < \frac{1}{NB}\sum \boldsymbol{x}_s.$$

It is clearly true for all k', $k - B < k' \leq k$, that the total load in transit is equal to the total system load minus the load at the nodes. Hence, if q is the column of $\boldsymbol{x}_k$ that contains the load at the nodes and in transit at time k', then

$$\begin{aligned}\sum(\boldsymbol{x}_t)^q &= \sum(\boldsymbol{x}_s)^q - \sum(\boldsymbol{x}_{n0})^q \\ &\geq \sum(\boldsymbol{x}_s)^q - N\max\{(\boldsymbol{x}_n)^q\} \\ &\geq \frac{1}{B}\sum\boldsymbol{x}_s - N\max\{\boldsymbol{x}_n\}.\end{aligned}$$

However,

$$|A|\max\{\boldsymbol{x}_t\} \geq \sum(\boldsymbol{x}_t)^q$$

for all k', $k - B < k' \leq k$. It follows that

$$\begin{aligned}\max\{\boldsymbol{x}_t\} &\geq \frac{1}{|A|}\left[\frac{1}{B}\sum\boldsymbol{x}_s - N\max\{\boldsymbol{x}_n\}\right] \\ &\geq \frac{1}{|A|}\left[\frac{1}{B}\sum\boldsymbol{x}_s - N\max\{\boldsymbol{x}_k\}\right].\end{aligned} \tag{4.40}$$

Because of the maximum network transit time, this $\max\{\boldsymbol{x}_t\}$ resulted from at the most $B-1$ load passes. Due to condition $(a)(ii)$ on $e_k \in g(\boldsymbol{x}_k)$ and the maximum network sensing time, each of these load passes must have been smaller than $\max\{\boldsymbol{x}_n\} - \min\{\boldsymbol{x}_n\}$. Hence,

$$\begin{aligned}\max\{\boldsymbol{x}_t\} &\leq (B-1)(\max\{\boldsymbol{x}_n\} - \min\{\boldsymbol{x}_n\}) \\ &\leq B(\max\{\boldsymbol{x}_k\} - \min\{\boldsymbol{x}_n\}).\end{aligned} \tag{4.41}$$

Equations (4.40) and (4.41) imply that

$$\max\{\boldsymbol{x}_k\} - \min\{\boldsymbol{x}_n\} \geq \frac{1}{B|A|}\left[\frac{1}{B}\sum\boldsymbol{x}_s - N\max\{\boldsymbol{x}_k\}\right] \tag{4.42}$$

and Equations (4.38) and (4.42) imply that

$$\begin{aligned}2\rho(\boldsymbol{x}_k, \mathcal{X}_b) &\geq \frac{1}{B|A|}\left[\frac{1}{B}\sum\boldsymbol{x}_s - N\max\{\boldsymbol{x}_k\}\right] \\ \frac{2B|A|}{N}\rho(\boldsymbol{x}_k, \mathcal{X}_b) &\geq \frac{1}{BN}\sum\boldsymbol{x}_s - \max\{\boldsymbol{x}_k\}.\end{aligned} \tag{4.43}$$

From Equation (4.38), it is clear that $2\rho(\boldsymbol{x}_k, \mathcal{X}_b) + \min\{\boldsymbol{x}_n\} \geq \max\{\boldsymbol{x}_k\}$. Hence, from Equation (4.43), it is clear that

$$\begin{aligned}\frac{2B|A|}{N}\rho(\boldsymbol{x}_k, \mathcal{X}_b) &\geq \frac{1}{BN}\sum\boldsymbol{x}_s - 2\rho(\boldsymbol{x}_k, \mathcal{X}_b) - \min\{\boldsymbol{x}_n\} \\ \left[\frac{2B|A|}{N} + 2\right]\rho(\boldsymbol{x}_k, \mathcal{X}_b) &\geq \frac{1}{BN}\sum\boldsymbol{x}_s - \min\{\boldsymbol{x}_n\} = V(\boldsymbol{x}_k).\end{aligned} \tag{4.44}$$

Because Equations (4.39) and (4.44) both bound $V(\boldsymbol{x}_k)$ from above, we can claim that $V(\boldsymbol{x}_k)$ is always bounded from above by the greater of the two bounds. Therefore, it is always true that

$$V(\boldsymbol{x}_k) \leq \left[2 + \frac{2B|A|}{N}\right] \rho(\boldsymbol{x}_k, \mathcal{X}_b) \tag{4.45}$$

so that condition (*ii*) of Theorem 3.8 is satisfied.

We now demonstrate that condition (*i*) of Theorem 3.8 is satisfied by our choice of $\rho(\boldsymbol{x}_k, \mathcal{X}_b)$ and $V(\boldsymbol{x}_k)$.

Notice that

$$\sum \boldsymbol{x}_s \geq \max\left\{\sum \boldsymbol{x}_{no}, \sum \boldsymbol{x}_{n1}\right\}.$$

It follows, then, from Equation (4.36), that

$$V(\boldsymbol{x}_k) \geq \frac{1}{2NB} \sum \boldsymbol{x}_n - \min\{\boldsymbol{x}_n\}. \tag{4.46}$$

In an analogous manner to the nondelay case, $\sum \boldsymbol{x}_n$ is minimized in terms of $\max\{\boldsymbol{x}_n\}$ and $\min\{\boldsymbol{x}_n\}$ when exactly one element of $\boldsymbol{x}_n$ is equal to $\max\{\boldsymbol{x}_n\}$ and the remaining elements of $\boldsymbol{x}_n$ are equal to $\min\{\boldsymbol{x}_n\}$. From this analysis and Equation (4.46), we have that

$$\begin{aligned} V(\boldsymbol{x}_k) &\geq \frac{1}{2NB}[\max\{\boldsymbol{x}_n\} + (2NB - 1)\min\{\boldsymbol{x}_n\}] - \min\{\boldsymbol{x}_n\} \\ &= \frac{1}{2NB}[\max\{\boldsymbol{x}_n\} - \min\{\boldsymbol{x}_n\}]. \end{aligned} \tag{4.47}$$

It is clear from Equations (4.34) and (4.35) that

$$\rho(\boldsymbol{x}_k, \mathcal{X}_b) \leq \max\{(\max\{\boldsymbol{x}_n\} - \min\{\boldsymbol{x}_n\}) \;,\; \max\{\boldsymbol{x}_t\}\}. \tag{4.48}$$

We must consider two cases. First, consider $\max\{\boldsymbol{x}_n\} - \min\{\boldsymbol{x}_n\} \geq \max\{\boldsymbol{x}_t\}$. Then, according to Equation (4.48),

$$\rho(\boldsymbol{x}_k, \mathcal{X}_b) \leq \max\{\boldsymbol{x}_n\} - \min\{\boldsymbol{x}_n\}. \tag{4.49}$$

Equations (4.49) and (4.47) yield

$$V(\boldsymbol{x}_k) \geq \frac{1}{2NB}\rho(\boldsymbol{x}_k, \mathcal{X}_b). \tag{4.50}$$

Now, consider $\max\{\boldsymbol{x}_t\} > \max\{\boldsymbol{x}_n\} - \min\{\boldsymbol{x}_n\}$, so that

$$\rho(\boldsymbol{x}_k, \mathcal{X}_b) \leq \max\{\boldsymbol{x}_t\}. \tag{4.51}$$

As before, the maximum load in transit at times $k - B + 1, \ldots, k - 1, k$, is the sum of at most $B - 1$ load passes, each of which must have been smaller than $\max\{\boldsymbol{x}_n\} - \min\{\boldsymbol{x}_n\}$. Hence,

$$\max\{\boldsymbol{x}_t\} \leq B[\max\{\boldsymbol{x}_n\} - \min\{\boldsymbol{x}_n\}]. \tag{4.52}$$

Using a slight manipulation of Equations (4.51) and (4.52), along with Equation (4.47) yields

$$\begin{aligned}\frac{1}{B}\rho(\boldsymbol{x}_k, \mathcal{X}_b) &\le \max\{\boldsymbol{x}_n\} - \min\{\boldsymbol{x}_n\} \\ \frac{1}{2NB^2}\rho(\boldsymbol{x}_k, \mathcal{X}_b) &\le \frac{1}{2NB}[\max\{\boldsymbol{x}_n\} - \min\{\boldsymbol{x}_n\}] \le V(\boldsymbol{x}_k). \qquad (4.53)\end{aligned}$$

Because Equations (4.50) and (4.53) both bound $V(\boldsymbol{x}_k)$ from below, we can claim that $V(\boldsymbol{x}_k)$ is always bounded from below by the lesser of the two bounds. Therefore, it is always true that

$$V(\boldsymbol{x}_k) \ge \frac{1}{2NB^2}\rho(\boldsymbol{x}_k, \mathcal{X}_b)$$

so that condition (i) of Theorem 3.8 is satisfied. Hence condition (i) of Theorem 3.1 is satisfied.

For the final condition of Theorem 3.1, we must show that $V(X(\boldsymbol{x}_0, E_k, k))$ is a nonincreasing function for $k \ge 0$ and all E_k such that $E_k E \in \boldsymbol{E}_B(\boldsymbol{x}_0)$. To see that this is the case, notice that once $\boldsymbol{x}_0$ is specified, $V(\boldsymbol{x}_k)$ varies only as $\min\{\boldsymbol{x}_n\}$ varies. Clearly, what we must show is that $\min\{\boldsymbol{x}_n\}$ is nondecreasing as a function of k. According to condition $(a)(ii)$ on $e_k \in g(\boldsymbol{x}_k)$, if $e_{\alpha(i)}^{i \to p(i)} \in e_k$ and $q(i) = \{j : j \in p(i) \text{ and } x_i \ge x_j^i\}$ then $x_i' \ge x_j^i$ for all $j \in q(i)$. In words, no node can pass so much load that its load level drops below its pre-pass perception of the load level of any node that it passed to. Therefore, because we are assured that $x_j^i \ge \min\{\boldsymbol{x}_n\}$ for all $i \in L$ and $j \in p(i)$, $x_i' \ge \min\{\boldsymbol{x}_n\}$ for all $i \in L$. Hence, $\min\{\boldsymbol{x}_n\}$ is a nondecreasing function of k. Thus, condition (ii) of Theorem 3.1 is satisfied and $\mathcal{X}_b$ is stable in the sense of Lyapunov with respect to $\boldsymbol{E}_B$.

In order to show that $\mathcal{X}_b$ is asymptotically stable in the large with respect to $\boldsymbol{E}_B$, we must show that for all $\boldsymbol{x}_0 \notin \mathcal{X}_b$ and all E_k, such that $E_k E \in \boldsymbol{E}_B(\boldsymbol{x}_0)$,

$$V(X(\boldsymbol{x}_0, E_k, k)) \longrightarrow 0 \quad \text{as} \quad k \longrightarrow \infty. \qquad (4.54)$$

If $\boldsymbol{x}_k \notin \mathcal{X}_b$, then $\boldsymbol{x}_{k+1}$ will represent a change of the load levels of all of the nodes included in some nonempty subset of L. Any change in the load of node $i \in L$ that is not positive must be due to the passing of load by node i at time k. In fact, from conditions $(a)(ii)$ and $(a)(iii)$ on $e_k \in g(\boldsymbol{x}_k)$, we have that, if $e_{\alpha(i)}^{i \to p(i)} \in e_k$, then

$$\begin{aligned}x_i' &\ge x_j^i + \gamma_{ij}[x_i - x_j^i] \text{ for some } j \text{ such that } \alpha_j(i) > 0 \\ &\ge \min\{\boldsymbol{x}_n\} + \gamma[x_i - \min\{\boldsymbol{x}_n\}]. \qquad (4.55)\end{aligned}$$

where $\gamma = \min_{(i,j) \in A}\{\gamma_{ij}\}$. Notice that Equation (4.55) is valid, even if $e_{\alpha(i)}^{i \to p(i)} \notin e_k$ or $\alpha(i) = (0, 0, \ldots, 0)$. Thus, for any time $k' > k$, $x_i(k') \ge \min\{\boldsymbol{x}_n\}$. Again notice that $\min\{\boldsymbol{x}_n\}$ is a nondecreasing function of k.

The question now becomes whether or not the passing of load is guaranteed to increase $\min\{\boldsymbol{x}_n\}$. Employing Equation (4.55) and the fact that $\min\{\boldsymbol{x}_n\}$ is a nondecreasing function of k, we will use induction to show that

$$x_i(k+l) \geq \min\{\boldsymbol{x}_n\} + (\gamma)^l[x_i - \min\{\boldsymbol{x}_n\}] \text{ for all } i \in L. \tag{4.56}$$

The case of $l = 1$ is simply Equation (4.55). Assume that Equation (4.56) is valid for some general l. Then, from Equations (4.55) and (4.56),

$$\begin{aligned} x_i(k+l+1) &\geq \min\{\boldsymbol{x}_n(k+l)\} + \gamma[x_i(k+l) - \min\{\boldsymbol{x}_n(k+l)\}] \\ &\geq \min\{\boldsymbol{x}_n\} + \gamma[x_i(k+l) - \min\{\boldsymbol{x}_n\}] \\ &\geq \min\{\boldsymbol{x}_n\} + \gamma\left[\min\{\boldsymbol{x}_n\} + (\gamma)^l[x_i - \min\{\boldsymbol{x}_n\}] - \min\{\boldsymbol{x}_n\}\right] \\ &= \min\{\boldsymbol{x}_n\} + (\gamma)^{l+1}[x_i - \min\{\boldsymbol{x}_n\}] \text{ for all } i \in L. \end{aligned}$$

Thus, we have shown that Equation (4.56) is valid in general.

Fix a time k, such that $\boldsymbol{x}_k \notin \mathcal{X}_b$. If $x_i > \min\{\boldsymbol{x}_n\}$ for all $i \in L$, then

$$x_i(k+m) \geq \min\{\boldsymbol{x}_n\} + (\gamma)^{2B-1}[x_i - \min\{\boldsymbol{x}_n\}] \tag{4.57}$$

for all $i \in L$ and $m \in \{1, 2, \ldots, 2B-1\}$. From Equation (4.57) and the definition of the state, it is clear that

$$\begin{aligned} \min\{\boldsymbol{x}_n(k+2B-1)\} &\geq \min\{\boldsymbol{x}_n\} + (\gamma)^{2B-1}[\min_i\{x_i\} - \min\{\boldsymbol{x}_n\}] \\ &> \min\{\boldsymbol{x}_n\}. \end{aligned} \tag{4.58}$$

Let $L_* \subset L$ be the set of all i, such that $x_i = \min\{\boldsymbol{x}_n\}$. It is possible that $|L_*| > 0$. Because $x_i(k') \geq \min\{\boldsymbol{x}_k\}$ for all $k' > k$, if $x_i = \min\{\boldsymbol{x}_n\}$, then $x_i(k-m) = \min\{\boldsymbol{x}_n\}$ for all $m \in \{1, 2, \ldots, 2B-1\}$. Thus, for any two nodes i and j such that $(i,j) \in A$ and $j \in L_*$, $x_j^i = \min\{\boldsymbol{x}_k\}$. According to the restrictions imposed on valid event strings by $\boldsymbol{E}_B$, there must be times k' and k'', $k \leq k' \leq k''$, $k'' < k + 2B$, such that $e_{\alpha_j(i)}^{i \to p(i)} \in e_{k'}$ and $e_{\alpha_j(i)}^{j \leftarrow i} \in e_{k''}$ for some $i \in L$ and $j \in L_*$ such that $(i,j) \in A$. Because $|L_*| < N$, the above passing and receiving scenario may have to transpire $N-1$ times, to ensure that $x_j(k_1) > \min\{\boldsymbol{x}_n\}$ for all $j \in L_*$ and for some $k_1 \geq k$. It is apparent that for $k_1 = k + 2NB$, $x_j(k_1) > \min\{\boldsymbol{x}_n\}$ for all $j \in L_*$. Let $L^* \subset L$ such that $L^* \bigcup L_* = L$ and $L^* \bigcap L_* = \phi$. From Equation (4.56),

$$\min_{i \in L^*}\{x_i(k_1)\} \geq \min\{\boldsymbol{x}_n\} + (\gamma)^{2NB}[\min_{i \in L^*}\{x_i\} - \min\{\boldsymbol{x}_n\}].$$

For any $j \in L_*$ that receives load at time $k'' < k_1$ that was passed at time $k' \geq k$, we have from the above equation, the fact that $x_j^i(k') = \min\{\boldsymbol{x}_n\}$, and Equation (4.56) that

$$\begin{aligned} x_j(k'') &\geq x_j(k''-1) + \gamma[\min_{i \in L^*}\{x_i(k')\} - \min\{\boldsymbol{x}_n\}] \\ &\geq \min\{\boldsymbol{x}_n\} + \gamma[\min_{i \in L^*}\{x_i(k')\} - \min\{\boldsymbol{x}_n\}] \end{aligned}$$

$$\begin{aligned} x_j(k'') &\geq \min\{\boldsymbol{x}_n\} \\ &+ \gamma[\min\{\boldsymbol{x}_n\} + (\gamma)^{k'-k}[\min_{i\in L^*}\{x_i\} - \min\{\boldsymbol{x}_n\}] - \min\{\boldsymbol{x}_n\}] \\ &\geq \min\{\boldsymbol{x}_n\} + (\gamma)^{k'-k+1}[\min_{i\in L^*}\{x_i\} - \min\{\boldsymbol{x}_n\}] \end{aligned}$$

From Equation (4.56),

$$\begin{aligned} x_j(k_1) &\geq \min\{\boldsymbol{x}_n\} + (\gamma)^{k_1-k''}[x_j(k'') - \min\{\boldsymbol{x}_n\}] \\ &\geq \min\{\boldsymbol{x}_n\} + (\gamma)^{k_1-k''}[\min\{\boldsymbol{x}_n\} + \\ &\quad (\gamma)^{k'-k+1}[\min_{i\in L^*}\{x_i\} - \min\{\boldsymbol{x}_n\}] - \min\{\boldsymbol{x}_n\}] \\ &\geq \min\{\boldsymbol{x}_n\} + (\gamma)^{2NB}[\min_{i\in L^*}\{x_i\} - \min\{\boldsymbol{x}_n\}]. \end{aligned}$$

Therefore,

$$\min_i\{x_i(k+2NB)\} \geq \min\{\boldsymbol{x}_n\} + (\gamma)^{2NB}[\min_{i\in L^*}\{x_i\} - \min\{\boldsymbol{x}_n\}]$$

and

$$\begin{aligned} \min\{\boldsymbol{x}_n(k+2(N+1)B)\} &\geq \min\{\boldsymbol{x}_n\} + (\gamma)^{2NB}[\min_{i\in L^*}\{x_i\} - \min\{\boldsymbol{x}_n\}] \\ &\geq \min\{\boldsymbol{x}_n\}. \end{aligned} \tag{4.59}$$

Equations (4.58) and (4.59) and the definition of $V(\boldsymbol{x}_k)$ imply that

$$V(\boldsymbol{x}_k) - V(\boldsymbol{x}_{k+2(N+1)B}) \geq (\gamma)^{2NB}[\min_{i\in L^*}\{x_i\} - \min\{\boldsymbol{x}_n\}] > 0. \tag{4.60}$$

Therefore, Equation (4.54) holds and $\mathcal{X}_b$ is asymptotically stable in the large with respect to $\boldsymbol{E}_B$. ■

4.4.2 Exponential Convergence to a Balanced State

We employ an exponential stability theorem to prove that $\rho(\boldsymbol{x}_k, \mathcal{X}_b)$ is bounded from above by an exponential $\zeta e^{-\alpha k}\rho(\boldsymbol{x}_0, \mathcal{X}_b)$ for some $\alpha > 0$ and $\zeta > 0$.

Theorem 4.9 (Continuous Load, Delays, Exponential Stability): *For the load processor network with delays as described above, the invariant set $\mathcal{X}_b$ is exponentially stable in the large with respect to $\boldsymbol{E}_B$.*

Proof: Lemmas 1, 2, and 3 are adaptations of lemmas from the proof in [100]. Fix processor i and time k. For any $j \in p(i)$ and any time $k' > k$, we will say that system condition $E_j(k')$ occurs if:

$$(i) \qquad x_j^i(k') < \min\{\boldsymbol{x}_n\} + \frac{\gamma}{2}\gamma^{k'-k}[x_i - \min\{\boldsymbol{x}_n\}] \tag{4.61}$$

$$(ii) \qquad e_{\alpha(i)}^{i\to p(i)} \in e_{k'},\ \alpha_j(i) \geq \gamma[x_i(k') - x_j^i(k')]. \tag{4.62}$$

Lemma 1: *If* $j \in p(i)$, $k_1 > k$, $e_{\alpha(i)}^{i \to p(i)} \in e_{k_1}$, *and* $E_j(k_1)$ *occurs, then* $E_j(\bar{k})$ *does not occur for* $\bar{k} \geq k_1 + 2B$.

Proof: Suppose $k_1 \geq k$, $e_{\alpha(i)}^{i \to p(i)} \in e_{k_1}$, and $E_j(k_1)$ occurs. From Equation (4.56),

$$x_i(k_1) \geq \min\{\boldsymbol{x}_n\} + \gamma^{k_1 - k}[x_i - \min\{\boldsymbol{x}_n\}]. \tag{4.63}$$

Subtracting Equation (4.61) with $k' = k_1$ from Equation (4.63) yields

$$\begin{aligned} x_i(k_1) - x_j^i(k_1) &\geq (1 - \frac{\gamma}{2})\gamma^{k_1 - k}[x_i - \min\{\boldsymbol{x}_n\}] \\ &\geq \frac{1}{2}\gamma^{k_1 - k}[x_i - \min\{\boldsymbol{x}_n\}]. \end{aligned}$$

If we let $k' = k_1$, Equation (4.62) yields

$$\begin{aligned} \alpha_j(i) &\geq \gamma[x_i(k_1) - x_j^i(k_1)] \\ &\geq \frac{\gamma}{2}\gamma^{k_1 - k}[x_i - \min\{\boldsymbol{x}_n\}]. \end{aligned} \tag{4.64}$$

According to the restrictions placed on valid event strings by $\boldsymbol{E}_a$, processor j will receive load $\alpha_j(i)$ at some time k_2, $k_1 \leq k_2 < k_1 + B$. Hence,

$$\begin{aligned} x_j(k_2 + 1) &\geq x_j(k_2) - \zeta + \alpha_j(i) \\ &\geq \min\{\boldsymbol{x}_n(k_2)\} + \alpha_j(i), \end{aligned}$$

where ζ is the total load (which may be zero) passed by processor j at time k_2. Using Equation (4.64) this becomes

$$\begin{aligned} x_j(k_2 + 1) &\geq \min\{\boldsymbol{x}_n\} + \alpha_j(i) \\ &\geq \min\{\boldsymbol{x}_n\} + \frac{\gamma}{2}\gamma^{k_1 - k}[x_i - \min\{\boldsymbol{x}_n\}] \\ &\geq \min\{\boldsymbol{x}_n\} + \frac{\gamma}{2}\gamma^{k_2 + 1 - k}[x_i - \min\{\boldsymbol{x}_n\}]. \end{aligned}$$

Using Equation (4.56), it follows that for all $k_3 > k_2 + 1$,

$$\begin{aligned} x_j(k_3) &\geq \min\{\boldsymbol{x}_n\} + \gamma^{k_3 - k_2 - 1}[x_j(k_2 + 1) - \min\{\boldsymbol{x}_n\}] \\ &\geq \min\{\boldsymbol{x}_n\} + \gamma^{k_3 - k_2 - 1}[\min\{\boldsymbol{x}_n\} + \\ &\quad \frac{\gamma}{2}\gamma^{k_2 + 1 - k}[x_i - \min\{\boldsymbol{x}_n\}] - \min\{\boldsymbol{x}_n\}] \qquad (4.65) \\ &\geq \min\{\boldsymbol{x}_n\} + \gamma^{k_3 - k}[x_i - \min\{\boldsymbol{x}_n\}]. \qquad (4.66) \end{aligned}$$

Because $k_2 < k_1 + B$, Equation (4.66) is valid for all $k_3 \geq k_1 + B$.

Let $k_4 \geq k_1 + 2B$, such that $e_{\alpha(i)}^{i \to p(i)} \in e_{k_4}$. According to the maximum system sensing time, there is some time k_5, $k_4 \geq k_5 > k_4 - B$, such that $x_j^i(k_4) = x_j(k_5)$. Equation (4.66) is valid at time k_5 and yields

$$\begin{aligned} x_j^i(k_4) - \min\{\boldsymbol{x}_n\} = x_j(k_5) - \min\{\boldsymbol{x}_n\} &\geq \frac{\gamma}{2}\gamma^{k_5 - k}[x_i - \min\{\boldsymbol{x}_n\}] \\ &\geq \frac{\gamma}{2}\gamma^{k_4 - k}[x_i - \min\{\boldsymbol{x}_n\}]. \end{aligned}$$

Therefore, Equation (4.61) does not hold at k_4 and $E_j(k_4)$ does not occur. ■

Lemma 2: *There exists some $\eta > 0$ such that for any $i \in L$, $k \geq 0$, $j \in p(i)$ and any $k' \geq k + 3NB$, we have*

$$x_j(k') \geq \min\{\boldsymbol{x}_n\} + \eta\gamma^{k'-k}[x_i - \min\{\boldsymbol{x}_n\}].$$

Proof: Fix i and k. Let $k_1, \ldots, k_N$ be times such that $e_{\alpha(i)}^{i \to p(i)} \in e_{k_m}$ and $k_{m-1}+2B < k_m \leq k_{m-1}+3B$ for all $m \in \{1, 2, \ldots, N\}$. According to Lemma 1, if $j \in p(i)$ and $m \neq l$, then $E_j(k_m)$ and $E_j(k_l)$ cannot both occur. Thus, there is some k_m, $m \in \{1, 2, \ldots, N\}$, such that $E_j(k_m)$ does not occur for any $j \in p(i)$. According to condition $(a)(iii)$ on $e_k \in g(\boldsymbol{x}_k)$, Equation (4.62) must be valid for some j^* such that $x_{j^*}^i(k_m) \leq x_j^i(k_m)$ for all $j \in p(i)$. Because $E_j(k_m)$ does not occur, Equation (4.61) is violated for $j = j^*$. It follows that, for all $j \in p(i)$,

$$x_j^i(k_m) \geq \min\{\boldsymbol{x}_n\} + \frac{\gamma}{2}\gamma^{k_m-k}[x_i - \min\{\boldsymbol{x}_n\}]. \tag{4.67}$$

According to the maximum system sensing time, there is some time k_1, $k_m \geq k_1 > k_m - B$, such that $x_j^i(k_m) = x_j(k_1)$. For any k_2, $k_2 \geq k + 3NB$, we have $k_2 \geq k_m \geq k_1$, and Equation (4.56) yields

$$x_j(k_2) \geq \min\{\boldsymbol{x}_n\} + \gamma^{k_2-k_1}[x_j(k_1) - \min\{\boldsymbol{x}_n\}].$$

Realizing that $x_j(k_1) = x_j^i(k_m)$, we employ Equation (4.67) to conclude that

$$\begin{aligned}
x_j(k_2) &\geq \min\{\boldsymbol{x}_n\} + \gamma^{k_2-k_1}[\min\{\boldsymbol{x}_n\} + \\
&\quad \frac{\gamma}{2}\gamma^{k_m-k}[x_i - \min\{\boldsymbol{x}_n\}] - \min\{\boldsymbol{x}_n\}] \\
&\geq \min\{\boldsymbol{x}_n\} + \frac{\gamma}{2}\gamma^{(k_2-k)+(k_m-k_1)}[x_i - \min\{\boldsymbol{x}_n\}]] \\
&\geq \min\{\boldsymbol{x}_n\} + \frac{\gamma}{2}\gamma^{k_2-k}\gamma^B[x_i - \min\{\boldsymbol{x}_n\}]].
\end{aligned}$$

This proves Lemma 1 with

$$\eta = \frac{\gamma^{B+1}}{2}.$$

■

Lemma 3: *For any $i \in L$, any $k \geq 0$, any $j \in L$ that can be reached from i by traversing l arcs, and for any $k' \geq k + 3lNB$, we have*

$$x_j(k') \geq \min\{\boldsymbol{x}_n\} + (\eta\gamma^{k'-k})^l[x_i - \min\{\boldsymbol{x}_n\}].$$

Proof: Lemma 2 establishes Lemma 3 for $l = 1$. Assume that Lemma 3 is true for every j at a distance of l from i. Assume m is at distance $l + 1$ from

i. Them $m \in p(j)$ for some j at a distance l from i. It follows, from our inductive hypothesis, that

$$x_j(k + 3lNB) \geq \min\{\boldsymbol{x}_n\} + (\eta\gamma^{3lNB})^l[x_i - \min\{\boldsymbol{x}_n\}].$$

If we apply Lemma 2 to processor $m \in p(j)$ at time $k_1 \geq k + 3lNB + 3NB$, we see that

$$\begin{aligned} x_m(k_1) &\geq \min\{\boldsymbol{x}_n\} + \eta\gamma^{k_1-k-3lNB}[x_j(k+3lNB) - \min\{\boldsymbol{x}_n\}] \\ &\geq \min\{\boldsymbol{x}_n\} + \eta\gamma^{k_1-k-3lNB}[\min\{\boldsymbol{x}_n\} + \\ &\quad (\eta\gamma^{3lNB})^l[x_i - \min\{\boldsymbol{x}_n\}] - \min\{\boldsymbol{x}_n\}] \\ &\geq \min\{\boldsymbol{x}_n\} + \eta\gamma^{k_1-k-3lNB}(\eta\gamma^{3lNB})^l[x_i - \min\{\boldsymbol{x}_n\}] \\ &\geq \min\{\boldsymbol{x}_n\} + \eta\gamma^{k_1-k}(\eta\gamma^{k_1-k})^l[x_i - \min\{\boldsymbol{x}_n\}] \\ &\geq \min\{\boldsymbol{x}_n\} + (\eta\gamma^{k_1-k})^{l+1}[x_i - \min\{\boldsymbol{x}_n\}]. \end{aligned}$$

Hence, the induction is complete and we have proven Lemma 3. ∎

Fix $i \in L$ and $k \geq 2B - 1$. Because every processor is at a distance of less than N from i, Lemma 3 yields

$$x_j(k') \geq \min\{\boldsymbol{x}_n\} + (\eta\gamma^{3N^2B+2B})^N[x_i - \min\{\boldsymbol{x}_n\}]$$

for all j, for all $k' \in [k + 3N^2B, k + 3N^2B + 2B]$. Hence,

$$\min\{\boldsymbol{x}_n(k + 3N^2B + 2B)\} \geq \min\{\boldsymbol{x}_n\} + (\eta\gamma^{3N^2B+2B})^N[x_i - \min\{\boldsymbol{x}_n\}].$$

This relation is true for all $i \in L$. Thus,

$$\begin{aligned} \min\{\boldsymbol{x}_n(k + 3N^2B + 2B)\} &\geq \min\{\boldsymbol{x}_n\} + (\eta\gamma^{3N^2B+2B})^N \cdot \\ &\quad [\max_i\{x_i\} - \min\{\boldsymbol{x}_n\}] \\ &\geq \min\{\boldsymbol{x}_n(k - 2B + 1)\} + (\eta\gamma^{3N^2B+2B})^N \cdot \\ &\quad [\max_i\{x_i\} - \min\{\boldsymbol{x}_n(k - 2B + 1\}]. \qquad (4.68) \end{aligned}$$

Invoking the definition of the state, it is clear that

$$\max\{\boldsymbol{x}_n\} = \max_i\{x_i(k - m)\} \text{ for some } m \in \{0, 1, \ldots, 2B - 1\}.$$

Equation (4.56) and the above equation imply that

$$\begin{aligned} \max_i\{x_i\} &\geq \min\{\boldsymbol{x}_n(k - m)\} + \gamma^m[\max_i\{x_i(k - m)\} - \min\{\boldsymbol{x}_n(k - m)\}] \\ &\geq \min\{\boldsymbol{x}_n(k - m)\} + \gamma^m[\max\{\boldsymbol{x}_n\} - \min\{\boldsymbol{x}_n(k - m)\}]. \end{aligned}$$

Manipulating further, we obtain

$$\begin{aligned} \max_i\{x_i\} - \min\{\boldsymbol{x}_n(k - m)\} &\geq \gamma^m[\max\{\boldsymbol{x}_n\} - \min\{\boldsymbol{x}_n(k - m)\}] \\ \max_i\{x_i\} - \min\{\boldsymbol{x}_n(k - 2B + 1)\} &\geq \gamma^{2B-1}[\max\{\boldsymbol{x}_n\} - \min\{\boldsymbol{x}_n(k - m)\}] \\ &\geq \gamma^{2B-1}[\max\{\boldsymbol{x}_n\} - \min\{\boldsymbol{x}_n\}]. \qquad (4.69) \end{aligned}$$

From Equations (4.68) and (4.69), it follows that

$$\begin{aligned}\min\{\boldsymbol{x}_n(k+3N^2B+2B)\} &\geq \min\{\boldsymbol{x}_n(k-2B+1)\} + \\ &\gamma^{2B-1}(\eta\gamma^{3N^2B+2B})^N[\max\{\boldsymbol{x}_n\} - \min\{\boldsymbol{x}_n\}]\end{aligned}$$

and

$$\min\{\boldsymbol{x}_n(k+3N^2B+2B)\} - \min\{\boldsymbol{x}_n(k-2B+1)\} \geq \gamma^{2B-1}(\eta\gamma^{3N^2B+2B})^N[\max\{\boldsymbol{x}_n\} - \min\{\boldsymbol{x}_n\}].$$

From Equation (4.36), it is apparent that

$$V(\boldsymbol{x}_{k+3N^2B+2B}) = \frac{1}{NB}\sum \boldsymbol{x}_s - \min\{\boldsymbol{x}_n(k+3N^2B+2B)\}$$

and

$$V(\boldsymbol{x}_{k-2B+1}) = \frac{1}{NB}\sum \boldsymbol{x}_s - \min\{\boldsymbol{x}_n(k-2B+1)\}.$$

Clearly, then,

$$\begin{aligned}V(\boldsymbol{x}_{k-2B+1}) - V(\boldsymbol{x}_{k+3N^2B+2B}) &= \min\{\boldsymbol{x}_n(k+3N^2B+2B)\} - \\ &\quad \min\{\boldsymbol{x}_n(k-2B+1)\} \\ &\geq \gamma^{2B-1}(\eta\gamma^{3N^2B+2B})^N \cdot \\ &\quad [\max\{\boldsymbol{x}_n\} - \min\{\boldsymbol{x}_n\}] \qquad (4.70)\end{aligned}$$

for all $k \geq 2B-1$. In the proof of asymptotic stability it is shown that

$$\max\{\boldsymbol{x}_n\} - \min\{\boldsymbol{x}_n\} \geq \frac{1}{B}\rho(\boldsymbol{x}_k, \mathcal{X}_b)$$

for all $\boldsymbol{x}_k \notin \mathcal{X}_b$. Hence, Equation (4.70) becomes

$$V(\boldsymbol{x}_{k-2B+1}) - V(\boldsymbol{x}_{k+3N^2B+2B}) \geq \frac{1}{B}\gamma^{2B-1}(\eta\gamma^{3N^2B+2B})^N\rho(\boldsymbol{x}_k, \mathcal{X}_b). \quad (4.71)$$

Lemma 4: *The closed, invariant set $\mathcal{X}_m \in \mathcal{X}$ is exponentially stable with respect to $\boldsymbol{E}_a$ if there exists a V defined on $S(\mathcal{X}_m; r)$, constants c_1, c_2, $c_3 > 0$, and constants D, D_1, such that $D > D_1$ and D, $D_1 \in N$, such that*

(*i*) $V(\boldsymbol{x}_{k+1}) \leq V(\boldsymbol{x}_k)$ for all $k \geq 0$

(*ii*) $c_1\rho(\boldsymbol{x}, \mathcal{X}_m) \leq V(\boldsymbol{x}) \leq c_2\rho(\boldsymbol{x}, \mathcal{X}_m)$ for all $\boldsymbol{x} \in S(\mathcal{X}_m; r)$

(*iii*) $V(\boldsymbol{x}_k) - V(\boldsymbol{x}_{k+D}) \geq c_3\rho(\boldsymbol{x}_{k+D_1}, \mathcal{X}_m)$ for all $k \geq 0$.

Proof: Conditions (i), (ii), and (iii) imply that

$$\begin{aligned} V(\boldsymbol{x}_k) - V(\boldsymbol{x}_{k+D}) &\geq c_3 \rho(\boldsymbol{x}_{k+D_1}, \mathcal{X}_m) \\ &\geq \frac{c_3}{c_2} V(\boldsymbol{x}_{k+D_1}) \\ &\geq \frac{c_3}{c_2} V(\boldsymbol{x}_{k+D}). \end{aligned} \tag{4.72}$$

We now show, via induction, that for all integers $m \geq 0$

$$V(\boldsymbol{x}_{mD}) \leq \left(1 + \frac{c_3}{c_2}\right)^{-m} V(\boldsymbol{x}_0) \tag{4.73}$$

Our induction hypothesis is that Equation (4.73) is valid for some $m \geq 0$. From Equation (4.72) and our induction hypothesis, it follows that

$$\begin{aligned} V(\boldsymbol{x}_{mD}) - V(\boldsymbol{x}_{(m+1)D}) &\geq \frac{c_3}{c_2} V(\boldsymbol{x}_{(m+1)D}) \\ V(\boldsymbol{x}_{mD}) &\geq \left(1 + \frac{c_3}{c_2}\right) V(\boldsymbol{x}_{(m+1)D}) \\ V(\boldsymbol{x}_{(m+1)D}) &\leq \left(1 + \frac{c_3}{c_2}\right)^{-1} V(\boldsymbol{x}_{mD}) \leq \left(1 + \frac{c_3}{c_2}\right)^{-(m+1)} V(\boldsymbol{x}_0) \end{aligned}$$

and Equation (4.73) is valid for $m + 1$. If we let $k = 0$ in Equation (4.72),

$$\begin{aligned} V(\boldsymbol{x}_0) &\geq \left(1 + \frac{c_3}{c_2}\right) V(\boldsymbol{x}_D) \\ V(\boldsymbol{x}_D) &\leq \left(1 + \frac{c_3}{c_2}\right)^{-1} V(\boldsymbol{x}_0) \end{aligned}$$

we see that Equation (4.73) is valid for $m = 1$. Therefore Equation (4.73) is valid for all $m \geq 0$ (Equation (4.73) is trivially satisfied for $m = 0$). Equation (4.73) and condition (i) imply that, for all k such that $(m-1)D \leq k \leq mD$,

$$V(\boldsymbol{x}_k) \leq \left(1 + \frac{c_3}{c_2}\right)^{-(m-1)} V(\boldsymbol{x}_0).$$

Because $(k/D) - 1 \leq m - 1$ for all k, such that $k \leq mD$, it follows, from the above equation, that

$$\begin{aligned} V(\boldsymbol{x}_k) &\leq \left(\frac{1}{1 + \frac{c_3}{c_2}}\right)^{(k/D)-1} V(\boldsymbol{x}_0) \\ V(\boldsymbol{x}_k) &\leq \left(1 + \frac{c_3}{c_2}\right) \left[\left(\frac{1}{1 + \frac{c_3}{c_2}}\right)^{1/D}\right]^k V(\boldsymbol{x}_0) \end{aligned} \tag{4.74}$$

From Equation (4.74) and condition (ii), we see that

$$\begin{aligned} c_1 \rho(\boldsymbol{x}_k, \mathcal{X}_m) &\leq \left(1 + \frac{c_3}{c_2}\right) \beta^k V(\boldsymbol{x}_0) \\ &\leq c_2 \left(1 + \frac{c_3}{c_2}\right) \beta^k \rho(\boldsymbol{x}_0, \mathcal{X}_m), \end{aligned}$$

where

$$\beta = \left(1 + \frac{c_3}{c_2}\right)^{-1/D} < 1.$$

Therefore, there is some $\alpha > 0$ such that $e^{-\alpha} \geq \beta$ and

$$\rho(\boldsymbol{x}_k, \mathcal{X}_m) \leq \frac{c_2}{c_1} \left(1 + \frac{c_3}{c_2}\right) e^{-\alpha k} \rho(\boldsymbol{x}_0, \mathcal{X}_m).$$

This completes the proof of Lemma 4. ■

Clearly, Equation (4.71) satisfies condition (iii) of Lemma 4. Conditions (i) and (ii) of Lemma 4 are satisfied in the proof of stability. This completes the proof of the theorem. ■

In the proof we show that

$$V(\boldsymbol{x}_{k-2B+1}) - V(\boldsymbol{x}_{k+3N^2B+2B}) \geq \frac{1}{B} \gamma^{2B-1} (\eta \gamma^{3N^2B+2B})^N \rho(\boldsymbol{x}_k, \mathcal{X}_b), \quad (4.75)$$

for $k \geq 2B-1$. Because Equation (4.75) is valid only for $k \geq 2B-1$, it should be apparent that $k = 2B-1$ in our model is equivalent to $k = 0$ in Lemma 4. Hence, what we have shown, via the proof, is for all $k \geq 2B-1$ that

$$\rho(\boldsymbol{x}_k, \mathcal{X}_b) \leq \zeta e^{-\alpha(k-2B+1)} \rho(\boldsymbol{x}_{2B-1}, \mathcal{X}_b)$$

for some $\alpha > 0$ and $\zeta > 0$. Of course, from the proof of asymptotic stability, we are assured that

$$\rho(\boldsymbol{x}_k, \mathcal{X}_b) \leq 2NB^2 \left(2 + \frac{2B|A|}{N}\right) \rho(\boldsymbol{x}_0, \mathcal{X}_b)$$

for all $0 \leq k < 2B-1$.

The value

$$\frac{1}{B} \gamma^{2B-1} (\eta \gamma^{3N^2B+2B})^N$$

from Equation (4.75) is directly related to the α from the exponential over-bounding function

$$\zeta e^{-\alpha(k-2B+1)} \rho(\boldsymbol{x}_{2B-1}, \mathcal{X}_b).$$

Thus, if speed of convergence is a design factor, then γ should be made as large as possible and N and B should be made as small as possible.

The condition $k' \geq k + 3NB$ in Lemma 2 is unnecessarily conservative. From the proof of Lemma 2, we see that the condition $k' \geq k + 3RB$, where $R = \max_i\{|p(i)|\} + 1$, is sufficient. Let S be the maximum number of arcs that must be spanned to reach any node $j \in L$ from any other node $i \in L$. Because every processor $i \in L$ is actually at a distance (in arcs) of S or less from every other processor $j \in L$, Equation (4.75) can be validly written as

$$V(\boldsymbol{x}_{k-2B+1}) - V(\boldsymbol{x}_{k+3RSB+2B}) \geq \frac{1}{B}\gamma^{2B-1}(\eta\gamma^{3RSB+2B})^S \rho(\boldsymbol{x}_k, \mathcal{X}_b).$$

Therefore, convergence can once again be accelerated by designing for RS^2 as small as possible.

Finally, note that the idea of virtual load works for the delay case similarly to the way in which it worked for the nondelay case.

4.5 SUMMARY

In this chapter we have performed an extensive case study of the load balancing system. We have studied stability in the sense of Lyapunov, asymptotic stability, and exponential stability for a variety of load balancing problems. The major topics covered in this chapter were the following:

- How to model all the different load balancing problems with the model G from Chapter 2.
- The invariant sets representing balanced conditions.
- Different types of load: discrete, continuous, and virtual.
- Different types of convergence: asymptotic, exponential.
- Global vs. local properties.
- Delays in sensing load levels and in passing load.
- How to choose a Lyapunov function for this application.

It is hoped that the proofs of this chapter will give the reader an idea of the type of analysis that must be done to prove that a system possesses stability properties.

4.6 FOR FURTHER STUDY

The load balancing systems that we examine are similar to and generalizations of those analyzed in [78, 77, 81] and in [100]. In [78, 77, 81], the load balancing system is very simple because the load is considered to exist only in blocks of unit size, the allowed interprocessor load exchanges are quite restricted, and any delays that exist in passing load and sensing load levels are ignored. The model in [100] assumes that load can be represented by a continuous variable and that delays exist in load passing and sensing. The model also allows for general load passing. The authors in [100] show that eventually the load will be perfectly balanced among the processors, and they suggest a proof for the "geometric convergence" of the network to a balanced state. This chapter is based on the work in [18, 15, 13], which was motivated by the work in [100].

The authors in [72] study how to minimize the number of load transfers to achieve balancing by using global information about the load distribution in the system. In all the problems considered in this chapter, the load processors only use local information, balancing proceeds in an asynchronous fashion, and we do not consider trying to minimize the number of load transfers to achieve balancing. Also note that the authors in [6] consider load balancing in a stochastic framework (our framework only admits deterministic load balancing problems).

4.7 PROBLEMS

Problem 4.1 (Stability Analysis of a Three Processor Load Balancing System): Suppose that you are given a load balancing system with three load processors, each of which is connected to all the others (i.e., in a ring).

(a) Suppose that you have discrete load with no delays. Provide an invariant set and prove that it is asymptotically stable and exponentially stable.

(b) Repeat (a) for continuous load and virtual load.

(c) Repeat (a) for continuous load with transfer and sensing delays.

Problem 4.2 (Simulation of a Load Balancing System):

(a) Simulate the load balancing system of Problem 4.1 for the discrete load case with no delays for a variety of initial load distributions.

(b) Repeat (a) for the case where there can be sensing and load transfer delays of up to two time units.

Problem 4.3 (Stability Proofs): Repeat each of the proofs in the chapter, filling in all the details.

Problem 4.4 (Stability Analysis of More General Load Balancing Systems)*: In this problem you will seek to extend the results of this chapter.

(a) Extend the case where there are delays in sensing and transfer to cover the possibility of discrete loads and virtual loads.

(b) In all of the load balancing problems considered in this chapter it is assumed that no new load arrives at the network for processing and that no load is processed while the load is being balanced. Certainly the results indicate that if load arrives/departs while it is being balanced, the system will continually seek to balance, but in general the systems will not possess the stability properties found in this chapter. What guarantees are needed to ensure boundedness of the buffers? With arrivals and departures how to you characterize a balanced load? Characterize and analyze stability properties of the general load balancing problem with arrivals/departures.

(c) Define a general load balancing problem in a stochastic framework and analyze stochastic stability properties of it.

5

Scheduling in Manufacturing Systems

5.1 OVERVIEW

Flexible manufacturing systems (FMS) are interconnected networks of machines that can be reconfigured in many ways to process many different types of parts. For instance, we may have a machine that does a variety of drilling operations, a machine that does certain types of assembly of components, and yet another that does some type of part inspection. For manufacturing one type of part it may be necessary to configure this set of machines so that, for instance, raw materials enter for assembly, then some drilling is done, then some more assembly (of a different type), then the part is passed back to the drilling machine, then it is inspected, then more assembly is done and then the finished part is output. Now, if the shop is given a different order, for a different type of part one can reconfigure the ordering and processing of the parts to construct a very different type of part on the same set of machines. Such flexibility in manufacturing has proven quite useful.

In this chapter we study a class of FMS that are composed of a machine, transport delays between machines, and a variety of elements, including a "stream modifier." We will focus on showing that the buffers that the parts wait in at each network element are bounded for a given FMS. In particular, we will show that a variety of scheduling policies for the machine are stable and will provide an efficient scheduling policy for the stream modifier. Moreover, we explain how these scheduling policies can be used to achieve a distributed real-time approach to scheduling FMS.

Next, we will show how to exploit more network level information in the scheduling of FMS (i.e., a centralized scheduling policy). In particular, we show that several policies that use information from the paths on which the parts will travel can ensure bounded buffers in the FMS. At the end of the chapter we will provide a battery of simulations to compare the centralized and distributed scheduling approaches.

The overall goal of this chapter is to solidify the boundedness concepts and analysis approaches that are covered in Chapter 3. The remainder of this book depends in no way on the topics covered in this chapter.

5.2 FMS SCHEDULING: A DISTRIBUTED APPROACH

An FMS is an interconnected network of machines and possibly other network elements. Such network elements may include bounded transport delays or stream modifiers that seek to hold and release parts when they are passed from one machine to another, in a way that will help to ensure that bounded network behavior is achieved. In this first section we show that a variety of network elements have bounded buffer levels if they are properly scheduled. In fact, we will focus on the local scheduling of network elements (i.e., how to schedule them without using information from other network elements). In this way we achieve a distributed approach to FMS scheduling in the sense that each network element will have its own scheduler and they will not need "global" information, that is, information about the other network elements.

We actually approach the distributed scheduling problem in a modular fashion by providing conditions for stability of individual machines in terms of constraints on the streams of parts that flow between machines. Then, we provide a stream modifier, which can change the stream constraints (to a certain extent), so that when the FMS machines are implemented with the stream modifiers, we will automatically get bounded buffers. It is for this reason that, in the following sections, we can focus on scheduling the network elements, and, if the reader is interested in connecting a particular topology, this can be done and, if the stream constraints can all be made to match properly (i.e., the output stream constraints match the input stream constraints of the machine that the parts are passed to), then the FMS will have bounded buffers.

5.2.1 Machine Model and Stream Constraints

The FMS considered here are networks of machines, each of which has the capability to process N different part types $i \in P$, where $P = \{1, 2, \ldots, N\}$. Parts that arrive at a machine and are awaiting servicing are held in buffers, and the buffer levels are denoted by $x_i(k)$, $i \in P$. Each machine can only process at some bounded rate one type of part at a time and, in general, a machine incurs a "set-up time" s_i (a bounded delay), when changing over to produce a new part type $i \in P$. The time during which a single part is being produced is called a "production run." We require the part flow to be

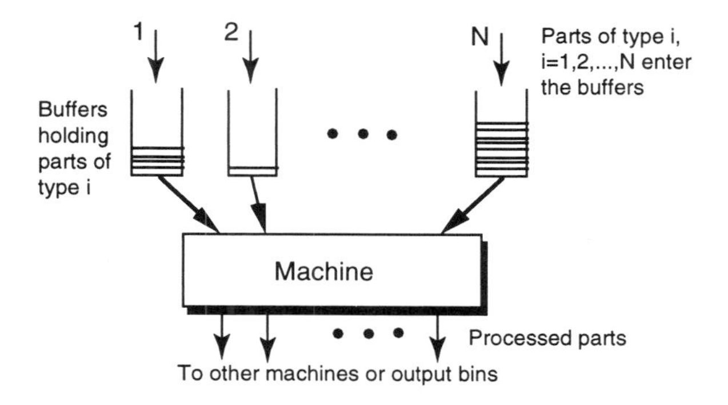

Fig. 5.1 Single machine for a flexible manufacturing system.

composed of discrete parts because this is the case in most practical FMS. In approaching the analysis and design of FMS, the critical elements are the part production schedules, or *scheduling policies*, for the component machines (see Figure 5.1).

Let $W = \{0, 1, 2, \ldots\}$ and let t_k, $k \in W$, be the *real time* corresponding to discrete time k. We will call the fixed length of real time between discrete times k and $k+1$ one *period*. As is the case with any discrete time model, the choice of period is important here. The single condition on the length of the period is that, when in the midst of a production run, the machine must produce at least one part per period. Because such a period can be chosen for any realistic machine (by choosing the period sufficiently large), the condition is not restrictive. While this assumption simplifies the notational logistics of the entire analysis, only in the case of the RPS policy does the analysis rely on this assumption in a substantial manner. If parts arrive at the machine at real times other than the t_k, $k \in W$, then our analysis is valid at the real time points t_k, $k \in W$, if we consider all parts that arrive at or depart from the machine in any real time interval $[t_k, t_{k+1})$, $k \in W$, to have done so at real time t_k. Let $\boldsymbol{x}_k = [x_1(k), x_2(k), \ldots, x_N(k)]^\top$, where $k \in W$. Let Z be any set of times such that $Z \subset W$ and for all k_1, $k_2 \in Z$, if $k_1 \leq k' \leq k_2$, then $k' \in Z$ and there is some $i \in P$, such that $x_i(k') > 0$. In other words, Z is any contiguous set of times during which the machine is not ever in a state of starvation (all buffers empty). For all of the analysis that follows, choose any such Z and, without loss of generality, assume that $\min Z = 0$. Let $A_i(k)$ be the (integer) number of parts of type $i \in P$ to arrive at the machine at time $k \in Z$, and let $D_i(k)$ be the (integer) number of parts of type $i \in P$ to depart from the machine at time $k \in Z$. At time $k \in Z$, the number of parts in buffer $i \in P$ is $x_i(k)$, and $x_i(k+1) = x_i(k) + A_i(k) - D_i(k)$. A part is considered to remain in its buffer until it exits the machine.

We define a function "ceil" such that $\text{ceil}(y) : \Re^+ \rightarrow W$ and $\text{ceil}(y) = \min\{k \in W : k \geq y\}$, for all $y \in \Re^+$ (i.e., ceil rounds its argument up to the nearest integer and only integer numbers of parts can arrive at or depart from machines). We define a function "floor" such that floor : $\Re^+ \to W$ and $\text{floor}(y) = \max\{k \in W : k \leq y\}$, for all $y \in \Re^+$ (i.e., floor rounds its argument down to the nearest integer). Let $F(k')$ be the number of production runs that have ended on or before time k'.

We will call any flow of parts into a machine an *input stream* and any flow of parts from a machine an *output stream*. We require that the input and output streams of the machine obey the following constraints:

(*i*) For all $k_1, k_2 \in Z$, $k_1 \leq k_2$, $i \in P$:

$$0 \leq \sum_{k=k_1}^{k_2} A_i(k) \leq \text{ceil}(a_i^{\text{in}}(k_2 - k_1 + 1) + b_i^{\text{in}}) . \tag{5.1}$$

(i.e., a_i^{in} is the maximum allowed rate, and b_i^{in} is the maximum allowed burstiness).

(*ii*) For all $k_1, k_2 \in Z$, such that k_1 and k_2 lie in the same production run

$$\text{floor}(\delta_j(k_2 - k_1 + 1)) \leq \sum_{k=k_1}^{k_2} D_j(k) \leq \text{ceil}(d_j(k_2 - k_1 + 1)) , \tag{5.2}$$

where $1 \leq \delta_j \leq d_j$ and j is the part type being produced (i.e., d_j and δ_j are rate constraints on output stream j, while the machine is producing parts of type j). The floor function in the output stream constraint is necessary to guarantee consistency (i.e., because δ_j and d_j need not be integers, it is possible that for some $k_1, k_2 \in W$, there is no $m \in W$, such that $\delta_j(k_2 - k_1 + 1) < m < d_j(k_2 - k_1 + 1)$, in which case there is no $\sum_{k=k_1}^{k_2} D_j(k)$ to satisfy Equation (5.2) with the floor function removed). The floor and ceil functions in constraints (*i*) and (*ii*) above are necessary, because a_i^{in}, δ_i, and d_i for $i \in P$ may be noninteger (e.g., if $a_1^{\text{in}} = 1.5$, we would like for three parts to be able to arrive at buffer 1 every two periods; however, if we remove the ceil function from constraint (*i*), only one part can arrive at buffer 1 each period).

Let

$$w_i = \frac{a_i^{\text{in}}}{\delta_i}$$

and $w = \sum_{i \in P} w_i$, where w_i is the ratio of maximum input rate and minimum output rate of buffer $i \in P$. It is intuitive that, to have any chance of maintaining bounded buffer levels, we must have $w < 1$ (this is often referred to as the *capacity condition*). In addition, for convenience let

$$u_i = \frac{b_i^{\text{in}}}{\delta_i}$$

and

$$u = \sum_{i \in P} u_i.$$

Notice that above we do not use the model G from Section 2.2 to represent the machine, and later for the network elements and FMS we do not use it either. Instead, we use more intuitive models, whose form is motivated by the manufacturing system application. The reader interested in using the model G can consult Problem 5.1 on page 143, where we model a machine very similar to the one here with the model G.

5.2.2 Clear-A-Fraction Policy

The CAF policy requires that, once a production run is begun, it will be continued until the buffer being processed is empty, then it chooses part type $j \in \{i \in P : x_i(k) \geq \epsilon \sum_{m=1}^{N} x_m(k)\}$, where $\epsilon \in (0, 1/N]$ to process. For notational simplicity let $\underline{\chi} := \min_i\{\chi_i\}$ and $\bar{\chi} := \max_i\{\chi_i\}$ for any variable χ_i, which is defined for all $i \in P$.

Theorem 5.1 (Clear-A-Fraction Policy): *When the CAF part servicing policy is used to control the above machine (with $w < 1$), it has buffer levels that are bounded for all $k \in W$ by*

$$\sum_{i \in P} x_i(k) \leq \bar{\delta}\left[\left(\sum_{i \in P} \frac{x_i(0)}{\delta_i} - \frac{\zeta}{1-\gamma}\right)\gamma^{F(k)} + \frac{\zeta}{1-\gamma} + \bar{s}w + u + \frac{N}{\underline{\delta}}\right], \tag{5.3}$$

where

$$\gamma = \max_i \left\{1 - \frac{\epsilon \underline{\delta}}{\delta_i}\left(\frac{1-w}{1-w_i}\right)\right\} \quad (0 < \gamma < 1); \tag{5.4}$$

$$\zeta = \max_i \left\{(u_i + s_i + 1)\left(\frac{w - w_i}{1 - w_i}\right) + (u - u_i) + \sum_{j \in P, j \neq i} \frac{1}{\delta_j}\right\} \tag{5.5}$$

and $x_i(k)$ is the number of parts in buffer $i \in P$ at time k, δ_i is the minimum processing rate of the machine, $F(k)$ is the number of production runs that have been completed by time k, and

$$u_i = \frac{b_i^{in}}{\delta_i}$$

with b_i^{in} the maximum allowed burstiness for the ith part type.

Proof: Choose $V(\boldsymbol{x}_k) = \sum_{i \in P} \frac{x_i(k)}{\delta_i}$. Because the following analysis is valid for any Z, the bounds obtained are valid for all $k \in W$ (the only k which are in the set W but not in any set Z are such that $x_i(k) = 0$ for all $i \in P$). We define the set of times $R = \{k_0, k_1, k_2, \ldots\} \subset Z$, $k_p < k_q$ if $p < q$, to include every time k' such that k' is the greatest time no longer than time k'', which immediately follows the end of any production run for some $k'' \in Z$. Notice that $k_0 = 0$. Let $j^*(k_p) \in P$, $k_p \in R$, denote the part type that is being setup for and processed by the machine between times k_p and k_{p+1}. We define $k_{p+1} - k_p \triangleq \Delta_p$. In order to bound Δ_p, we use an approach similar to that in [86], and after some manipulations, obtain

$$\Delta_p \leq \frac{\frac{x_{j^*(k_p)}(k_p)}{\delta_{j^*(k_p)}} + \frac{b^{\text{in}}_{j^*(k_p)}+1}{\delta_{j^*(k_p)}} + s_{j^*(k_p)} + 1}{1 - w_{j^*(k_p)}}. \tag{5.6}$$

We now bound $V(\boldsymbol{x}_{k_{p+1}})$ in terms of $V(\boldsymbol{x}_{k_p})$. Notice that $x_{j^*(k_p)}(k_{p+1}) = 0$ and $x_i(k_{p+1}) - x_i(k_p) \leq \text{ceil}(a_i^{\text{in}}\Delta_p + b_i^{\text{in}})$ for all $i \in P$, $i \neq j^*(k_p)$. From this, it follows that

$$\sum_{i \in P} \frac{x_i(k_{p+1})}{\delta_i} \leq \sum_{i \in P} \frac{x_i(k_p)}{\delta_i} - \frac{x_{j^*(k_p)}(k_p)}{\delta_{j^*(k_p)}} + \sum_{i \in P, i \neq j^*(k_p)} \frac{a_i^{\text{in}}\Delta_p + b_i^{\text{in}} + 1}{\delta_i}. \tag{5.7}$$

After some manipulations, we see that

$$\begin{aligned}
V(\boldsymbol{x}_{k_{p+1}}) &\leq V(\boldsymbol{x}_{k_p}) - \frac{x_{j^*(k_p)}(k_p)}{\delta_{j^*(k_p)}} + \sum_{i \in P, i \neq j^*(k_p)} \frac{a_i^{\text{in}}\Delta_p + b_i^{\text{in}} + 1}{\delta_i} \\
&\leq V(\boldsymbol{x}_{k_p}) - \frac{x_{j^*(k_p)}(k_p)}{\delta_{j^*(k_p)}} + \Delta_p(w - w_{j^*(k_p)}) + (u - u_{j^*(k_p)}) \\
&\quad + \sum_{i \in P, i \neq j^*(k_p)} \frac{1}{\delta_i} \\
&\leq V(\boldsymbol{x}_{k_p}) - \frac{x_{j^*(k_p)}(k_p)}{\delta_{j^*(k_p)}} \left(\frac{1 - w}{1 - w_{j^*(k_p)}} \right) + (u_{j^*(k_p)} + s_{j^*(k_p)} + 1) \cdot \\
&\quad \left(\frac{w - w_{j^*(k_p)}}{1 - w_{j^*(k_p)}} \right) + (u - u_{j^*(k_p)}) + \sum_{i \in P, i \neq j^*(k_p)} \frac{1}{\delta_i}
\end{aligned} \tag{5.8}$$

From the definition of the CAF policy, we see that $x_{j^*(k_p)}(k_p) \geq \epsilon\underline{\delta}V(\boldsymbol{x}_{k_p})$. Up to this point, this proof is similar to the CAF stability proof in [86], except that, in this proof, bounded input stream rate and burstiness constraints are allowed. If we define γ and ζ as in the statement of the theorem we see from Equation (5.8) that $V(\boldsymbol{x}_{k_{p+1}}) \leq \gamma V(\boldsymbol{x}_{k_p}) + \zeta$, for all $k_p \in R$ (notice that, by definition, $0 < \epsilon\underline{\delta}/\delta_i < 1$ for all $i \in P$ and $0 < \gamma < 1$). This is simply a difference inequality which, when solved, yields

$$V(\boldsymbol{x}_{k_p}) \leq \left(V(\boldsymbol{x}_0) - \frac{\zeta}{1 - \gamma} \right) \gamma^p + \frac{\zeta}{1 - \gamma} \triangleq G(\boldsymbol{x}_0, p). \tag{5.9}$$

Thus, we have bounded $V(\boldsymbol{x}_k)$ for all $k \in R$. Consider now the set of times S_p such that if $k \in Z$ and $k \in (k_p, k_{p+1})$, then $k \in S_p$. In Equation (5.9), we have found a bound $G(\boldsymbol{x}_0, p)$ for $V(\boldsymbol{x}_k)$, for $k = k_p$ and k_{p+1}. We now wish to bound $V(\boldsymbol{x}_k)$ for all $k \in S_p \cup \{k_p,\ k_{p+1}\}$. Clearly, the maximum of V over $S_p \cup \{k_p,\ k_{p+1}\}$ must occur at at one of the following times: k_p, $k_p + 1$ or at beginning of the production run that began before k_{p+1}, which we denote by $b(k_{p+1})$. We can bound the increase in V that occurs between times k_p and $b(k_{p+1})$ as

$$V(\boldsymbol{x}_{b(k_{p+1})}) - V(\boldsymbol{x}_{k_p}) \ \leq \ \sum_{i \in P} \frac{a_i^{\text{in}} s_i + b_i^{\text{in}} + 1}{\delta_i} \leq \bar{s}w + u + \frac{N}{\underline{\delta}} \ . \quad (5.10)$$

Hence, for all $k \in S_p \cup \{k_p, k_{p+1}\}$,

$$V(\boldsymbol{x}_k) \ \leq \ G(\boldsymbol{x}_0, p) + \bar{s}w + u + \frac{N}{\underline{\delta}},$$

and so for any $k \in Z$,

$$\begin{aligned} V(\boldsymbol{x}_k) \ &\leq \ G\left(\boldsymbol{x}_0, F(k)\right) + \bar{s}w + u + \frac{N}{\underline{\delta}} \\ &= \ \left(V(\boldsymbol{x}_0) - \frac{\zeta}{1-\gamma}\right)\gamma^{F(k)} + \frac{\zeta}{1-\gamma} + \bar{s}w + u + \frac{N}{\underline{\delta}} \ . \quad (5.11) \end{aligned}$$

Therefore, a bound on the buffer levels for all $k \in Z$, and, hence, for all $k \in W$, is

$$\begin{aligned} \frac{1}{\bar{\delta}} \sum_{i \in P} x_i(k) \ &\leq \ \sum_{i \in P} \frac{x_i(k)}{\delta_i} \leq \left(\sum_{i \in P} \frac{x_i(0)}{\delta_i} - \frac{\zeta}{1-\gamma}\right)\gamma^{F(k)} + \\ &\qquad \frac{\zeta}{1-\gamma} + \bar{s}w + u + \frac{N}{\underline{\delta}}, \end{aligned}$$

which completes the proof. ∎

Notice that, from Equation (5.3), a useful property of the machine buffer dynamics is apparent. If

$$\sum_{i \in P} \frac{x_i(0)}{\delta_i} > \frac{\zeta}{1-\gamma},$$

then the right side of Equation (5.3) must asymptotically *decrease* to

$$\bar{\delta}\left(\frac{\zeta}{1-\gamma} + \bar{s}w + u\right)$$

as $k \to \infty$, so that the sum of the buffer levels will get no larger than the bound at time $k = 0$. If

$$\sum_{i \in P} \frac{x_i(0)}{\delta_i} < \frac{\zeta}{1-\gamma},$$

then the right side of Equation (5.3) must asymptotically *increase* to

$$\bar{\delta}\left(\frac{\zeta}{1-\gamma}+\bar{s}w+u\right)$$

as $k \to \infty$. Hence, Equation (5.3) helps to characterize both the transient and steady state behavior of the machine. Notice that our proof is for a more general class of stream constraints for the parts arriving at and departing from the machine than in [86], and Equation (5.3) provides a more detailed characterization of the transient behavior of the machine. It is also possible to define a generalized CAF (GCAF), in which only a fixed fraction of the parts in a buffer at the beginning of a production run are cleared during that production run. Similar results hold for the GCAF policy. Finally, recall that the buffer bound established above is only valid at real times t_k. However, if the "intersample" input and output stream functions are known or can be bounded, then the above bound can be modified so that it is valid for all $t \in \Re^+$. If we know nothing of the "intersample" input and output stream functions, we can add d_i for all $i \in P$ to the buffer bounds, so that the bounds are valid for all $t \in \Re^+$.

5.2.3 Clear-Average-Oldest-Buffer Policy

We now introduce what we call the "clear the average oldest buffer" (CAOB) policy. Under this clearing policy, the buffer to be serviced is chosen according to a criterion based on the maximum number of periods that any of its parts have been waiting for service. For all $i \in P$ and for all $k \in W$, let $T_i(k)$ be the maximum number of periods that any part in buffer i has been waiting for service at time k and let $T(k) = [T_1(k), T_2(k), \ldots, T_N(k)]^t$. Let the CAOB policy choose at time k a part $j \in \{i \in P : T_i(k) \geq \epsilon \sum_{m=1}^{N} T_m(k)\}$ to process where $\epsilon \in (0, 1/N]$.

Theorem 5.2 (Clear-Average-Oldest-Buffer Policy): *If the CAOB part servicing policy is used to control the above machine and if $w_i < 1/N$ for all $i \in P$, then it has buffer levels that are bounded for all $k \in W$ by*

$$\begin{aligned}\sum_{i\in P} x_i(k) \quad &\leq \quad \bar{a}^{in}\left[\left(1-\gamma^{F(k-1)+1}\right)\frac{\zeta}{1-\gamma}+\bar{\eta}\right] \\ &\quad +\sum_{i\in P}\left(x_i(0)+b_i^{in}+1\right), \end{aligned} \tag{5.12}$$

where

$$\gamma \quad = \quad \max_i\left\{1-\epsilon\left(1-\frac{(N-1)a_i^{in}}{\delta_i(1-w_i)}\right)\right\} \qquad (0 < \gamma < 1) \tag{5.13}$$

$$\zeta \quad = \quad \max_i\left\{\frac{N-1}{1-w_i}\left(\frac{2b_i^{in}+1}{\delta_i}+s_i+1\right)\right\} \tag{5.14}$$

$$\eta_i = \max\{k \in W : \mathit{ceil}(a_i^{in}(k+1) + b_i^{in}) - \mathit{floor}(\delta_i(k+1-s_i)) > 0\}. \quad (5.15)$$

Proof: Choose

$$V(T(k)) = \sum_{i \in P} T_i(k).$$

Beginning in a similar way to the proof of Theorem 5.1 we know that Equation (5.6) holds for CAOB. Notice that $T_{j^*(k_{p+1})}(k_{p+1}) = 0$, $T_i(k_{p+1}) - T_i(k_p) \leq \Delta_p$ for all $i \in P$, $i \neq j^*(k_p)$, and $x_i(k) \leq a_i^{\text{in}}(T_i(k)) + b_i^{\text{in}} + 1$ for all $i \in P$, $k \in Z$. From this and from the definition of the CAOB policy, we see that

$$\sum_{i \in P} T_i(k_{p+1}) \leq \sum_{i \in P} T_i(k_p) - T_{j^*(k_p)}(k_p) + (N-1)\Delta_p \ .$$

Hence,

$$\begin{aligned} V(T(k_{p+1})) &\leq V(T(k_p)) - \epsilon V(T(k_p)) \left[1 - \frac{(N-1)a_{j^*(k_p)}^{\text{in}}}{\delta_{j^*(k_p)}(1 - w_{j^*(k_p)})} \right] \\ &\quad + \frac{N-1}{1 - w_{j^*(k_p)}} \left[\frac{2b_{j^*(k_p)}^{\text{in}} + 1}{\delta_{j^*(k_p)}} + s_{j^*(k_p)} + 1 \right] \\ &= V(T(k_p)) \left[1 - \epsilon \left(1 - \frac{(N-1)a_{j^*(k_p)}^{\text{in}}}{\delta_{j^*(k_p)}(1 - w_{j^*(k_p)})} \right) \right] \\ &\quad + \frac{N-1}{1 - w_{j^*(k_p)}} \left[\frac{2b_{j^*(k_p)}^{\text{in}} + 1}{\delta_{j^*(k_p)}} + s_{j^*(k_p)} + 1 \right] . \end{aligned} \quad (5.16)$$

If we define γ and ζ as in the statement of the theorem, we see from Equation (5.16) that

$$V(T(k_{p+1})) \leq \gamma V(T(k_p)) + \zeta,$$

for all $k_p \in R$. Notice that, by assumption, $0 < \gamma < 1$. Hence, we have

$$V(T(k_p)) \leq (1 - \gamma^p) \frac{\zeta}{1 - \gamma} \ \triangleq \ G(p).$$

(note that $V(T(k_0)) = 0$) and so we have bounded $V(T(k))$ for all $k \in R$.

Consider now the set of times S_p, such that if $k \in Z$ and $k \in (k_p, k_{p+1})$, then $k \in S_p$. We have found a bound $G(p)$ for $V(T(k))$ for $k = k_p$ and k_{p+1}. We now wish to bound $V(T(k))$ for all $k \in S_p \cup \{k_p,\ k_{p+1}\}$. To do this, we first must bound $T_{j^*(k_p)}(k') - T_{j^*(k_p)}(k_p)$ for all $k' \in S_p$. Consider the following expression:

$$\sum_{k=k_p-T_{j^*(k_p)}(k_p)}^{k_p-T_{j^*(k_p)}(k_p)+k'} \left[A_{j^*(k_p)}(k) - D_{j^*(k_p)}(k + T_{j^*(k_p)}(k_p))\right] , \qquad (5.17)$$

for all $k_p + k' \in S_p$. Expression (5.17) is the sum of all parts that arrived at buffer $j^*(k_p)$ at any time $k \in W$, $k_p - T_{j^*(k_p)}(k_p) \leq k \leq k_p - T_{j^*(k_p)}(k_p) + k'$ minus the sum of all parts which leave buffer $j^*(k_p)$ at any time $k \in S_p$, $k_p \leq k \leq k_p + k'$. If Expression (5.17) is positive for a given $k' \in W$, then more parts arrived at buffer $j^*(k_p)$ in the $k' + 1$ periods, beginning at time $k_p - T_{j^*(k_p)}(k_p)$ than have left buffer $j^*(k_p)$ in the $k' + 1$ periods beginning at time k_p. Hence, there is at least one part that arrived at buffer $j^*(k_p)$ on or before time $k_p - T_{j^*(k_p)}(k_p) + k'$, which remains in the buffer at time $k_p + k' + 1$ so that $T_{j^*(k_p)}(k_p + k') > T_{j^*(k_p)}(k_p)$. By the same reasoning, if Equation (5.17) is not positive, then $T_{j^*(k_p)}(k_p + k') \leq T_{j^*(k_p)}(k_p)$. It is clear that $T_{j^*(k_p)}(k_p + k') - T_{j^*(k_p)}(k_p) \leq k'$. We can bound Expression (5.17) by specifying that parts in buffer $j^*(k_p)$ be serviced as slowly as possible, so that

$$\begin{aligned} &\sum_{k=k_p-T_{j^*(k_p)}(k_p)}^{k_p-T_{j^*(k_p)}(k_p)+k'} \left[A_{j^*(k_p)}(k) - D_{j^*(k_p)}(k + T_{j^*(k_p)}(k_p))\right] && (5.18)\\ \leq\ & \text{ceil}(a^{\text{in}}_{j^*(k_p)}(k' + 1) + b^{\text{in}}_{j^*(k_p)}) \\ &-\text{floor}(\delta_{j^*(k_p)}(k' + 1 - s_{j^*(k_p)})) && (5.19)\end{aligned}$$

for all $k_p + k' \in S_p$. For all $i \in P$, let η_i be defined as given in the statement of Theorem 5.2. Notice that, because $a^{\text{in}}_i \leq \delta_i/N$ for all $i \in P$, and because we are inherently assuming that $b^{\text{in}}_i < \infty$ for all $i \in P$, $\eta_i < \infty$ for all $i \in P$. Then, for all $k_p + k' \in S_p$, $T_{j^*(k_p)}(k_p + k') - T_{j^*(k_p)}(k_p) \leq \eta_{j^*(k_p)}$. Hence, for all $k_p + k' \in S_p$, $V(T(k_p + k')) \leq G(p + 1) + \eta_{j^*(k_p)}$. Notice in the above bound that $G(p + 1)$ appears rather than $G(p)$. This is due to the fact that, for all $i \in P$, $i \neq j^*(k_p)$, $T_i(k_{p+1}) > T_i(k_p + k')$ for all $k_p + k' \in S_p$. Hence,

$$V(T(k)) \ \leq \ (1 - \gamma^{F(k-1)+1})\frac{\zeta}{1 - \gamma} + \bar{\eta} \qquad (5.20)$$

for all $k \in Z$, $k > 0$, and, hence, for all $k \in W$, $k > 0$. Because $x_i(k) \leq x_i(0) + a^{\text{in}}_i T_i(k) + b^{\text{in}}_i$ for all $k \in W$ and for all $i \in P$, it is clear that

$$\sum_{i \in P} x_i(k) \leq \sum_{i \in P} \left(x_i(0) + a^{\text{in}}_i T_i(k) + b^{\text{in}}_i + 1\right),$$

and

$$\sum_{i \in P} \left(x_i(0) + a^{\text{in}}_i T_i(k) + b^{\text{in}}_i + 1\right) \leq \bar{a}^{\text{in}} V(T(k)) + \sum_{i \in P} \left(x_i(0) + b^{\text{in}}_i + 1\right)$$

which with Equation (5.20), gives the final result for all $k \in W$, $k > 0$. ■

Notice that, in bounding $V(T(k))$ in Equation (5.20), that the bound increases from $\bar{\eta}$ to $\zeta/(1-\gamma)+\bar{\eta}$ as $k \to \infty$, so that we also characterize the transient properties of CAOB. A clear the oldest buffer (COB) policy is a special case of the CAOB policy; hence, the bounds above hold for this policy also. The COB policy is sometimes called the "first come first clear" (FCFC) policy since it will service the buffer that contains the part that arrived before all other parts in any of the other buffers. If parts tend to arrive at the machine such that a group of parts arriving at one buffer is followed by a group of parts arriving at another buffer, and so on (at a low enough frequency), then the CAOB policy will tend to behave like a FCFS policy.

5.2.4 Random Part Selection Policy

We now introduce what we call the "random part selection" (RPS) policy. Under this policy, the machine is free to choose any nonempty buffer to service at any time just so long as it never sits idle (i.e., it is either setting up for or processing a part at every instant).

Theorem 5.3 (Random Part Selection Policy: Case 1): *If $a_i^{in} \leq 1/N(\bar{s}+1)$ for all $i \in P$, and the RPS part servicing policy is used to control the above machine, then $\sum_{i=1}^N x_i(k) \leq b + N + 1 + \sum_{i=1}^N x_i(0)$, for all $k \in W$, where $b = \sum_{i=1}^N b_i^{in}$.*

Proof: Let $A(k)$ be defined so that $A(k) = \sum_{i=1}^N A_i(k)$, for all $k \in W$. In the worst case, in which the machine produces only a single part from any buffer before switching production to a different buffer, it is clear that it can take no longer than $\bar{s}+1$ periods to produce one part. If we let $D(k) = \max\{D_i(k) : i \in P\}$, we see that

$$\sum_{k=k_1}^{k_2} D(k) \geq \text{floor}\left(\frac{k_2 - k_1 + 1}{\bar{s}+1}\right),$$

for all $k_1, k_2 \in Z$, $k_1 \leq k_2$. From this, the definition of $A(k)$, and the assumption on a_i^{in}, $i \in P$, in the statement of the theorem, it is apparent that

$$\begin{aligned}
\sum_{k=0}^{k'} [A(k) - D(k)] &\leq \sum_{i=1}^{N} \left(\frac{k'+1}{N(\bar{s}+1)} + b_i^{\text{in}} + 1\right) \\
&\quad - \text{floor}\left(\frac{k'+1}{\bar{s}+1}\right) && (5.21) \\
&= \frac{k'+1}{\bar{s}+1} + b + N - \text{floor}\left(\frac{k'+1}{\bar{s}+1}\right) && (5.22) \\
&\leq b + N + 1\,, && (5.23)
\end{aligned}$$

and that

$$\sum_{k=0}^{k'} [A(k) - D(k)] = \sum_{i=1}^{N} x_i(k'+1) - \sum_{i=1}^{N} x_i(0),$$

for all $k', k'+1 \in Z$. Clearly, then,

$$\sum_{i=1}^{N} x_i(k'+1) \leq b + N + 1 + \sum_{i=1}^{N} x_i(0),$$

for all $k', k'+1 \in W$. ∎

Notice that, unlike the conditions on stability of the CAF policy, the condition for stability of the RPS policy does not depend on the processing speed of the machine (of course, we have required previously that the period length be chosen so that $1 \leq \delta_i \leq d_i$ for all $i \in P$). Rather, the condition simply limits the rates of the input streams of the machine. Intuitively, the RPS policy is stable because it is *persistent*, in that if there are parts in any machine buffer, it will always be either processing parts or setting up to process parts. The constraints on the rates of arrival of parts essentially ensure that the policy will switch from processing one buffer to another often enough so that no buffer can be ignored for too long. Several commonly used policies are special cases of the RPS policy, and, hence, are stable because they are persistent: (a) the first-come first-serve (FCFS) policy, (b) the priority policy (buffers are serviced in a fixed order, but empty buffers are skipped as in [75]), and fixed-time policy (nonempty buffers are serviced for a fixed amount of time). Moreover, policies studied in [42], such as the "earliest due date" policy, are special cases of RPS. It is interesting to note that, in [94], the author was able to show that FCFS is unstable for certain FMS topologies, where there are no setup times. The key to obtaining stability here is that, unlike in [94], we constrain the rates at which parts may be input to machines (so that, if applied to a network of machines, our results would require a stream modifier, like in the next section to achieve stable operation).

The stability conditions for the RPS policy in Theorem 5.3 are are not entirely satisfactory, because we cannot affect the input stream rate constraints by speeding up the machine, and this is contrary to our intuition. In light of this, we now reformulate the problem by altering the way in which we look at part arrivals and departures. First of all, it is necessary to redefine the period for this analysis. We choose the new period and constants $a_i^{\text{in}'}$, so that for every $i \in P$, there are at least $a_i^{\text{in}'}$ periods for each part that arrives at buffer i that is not attributable to the input stream burstiness. Similarly, choose δ_i', so that, when the machine is producing parts of type i, it outputs parts no slower than one part every δ_i' periods. As before, let s_i be the number of periods needed to set up for production on buffer $i \in P$. Assume that for all $i \in P$, $a_i^{\text{in}'}$ and δ_i' are integers (this assumption is not limiting, since we can choose the period to be as small as desired).

Theorem 5.4 (Random Part Selection Policy: Case 2): *If the RPS part servicing policy is used to control the above machine, and*

$$\max_i\{\delta_i' + s_i\} \sum_{i\in P} \frac{1}{a_i^{in'}} \leq 1,$$

then

$$\sum_{i=1}^{N} x_i(k) \leq b + N + 1 + \sum_{i=1}^{N} x_i(0),$$

for all $k \in W$, where $b = \sum_{i=1}^{N} b_i^{in}$.

Proof: Let $A(k) = \sum_{i\in P} A_i(k)$ and $D(k) = \sum_{i\in P} D_i(k)$, for all $k \in W$, so that

$$\begin{aligned}\sum_{k=k_1}^{k_2} A_i(k) &\leq \text{floor}\left(\frac{k_2 - k_1 + 1}{a_i^{\text{in}'}}\right) + 1 + b_i^{\text{in}} \\ &\leq \frac{k_2 - k_1 + 1}{a_i^{\text{in}'}} + 1 + b_i^{\text{in}}, \qquad (5.24)\end{aligned}$$

$$\sum_{k=k_1}^{k_2} A(k) \leq (k_2 - k_1 + 1) \sum_{i\in P} \frac{1}{a_i^{\text{in}'}} + N + \sum_{i\in P} b_i^{\text{in}}, \qquad (5.25)$$

and

$$\sum_{k=k_1}^{k_2} D(k) \geq \text{floor}\left(\frac{k_2 - k_1 + 1}{\max_i\{\delta_i' + s_i\}}\right) \geq \frac{k_2 - k_1 + 1}{\max_i\{\delta_i' + s_i\}} - 1,$$

for all $k_1, k_2 \in W$, $k_1 \leq k_2$. Notice that, as in the previous analysis of the RPS policy, we have identified the maximum number of periods per part serviced (i.e., as long as there are parts in any machine buffer, the machine must output at least one part every $\max_i\{\delta_i' + s_i\}$ periods). Clearly,

$$\begin{aligned}\sum_{k=0}^{k'} [A(k) - D(k)] &= \sum_{i\in P} (x_i(k'+1) - x_i(0)) \\ &\leq (k'+1)\left[\sum_{i\in P} \frac{1}{a_i^{\text{in}'}} - \frac{1}{\max_i\{\delta_i' + s_i\}}\right] + N + b + 1,\end{aligned}$$

for all $k' \in W$. Now, by the assumption in the theorem, we see that

$$\sum_{i\in P} (x_i(k+1) - x_i(0)) \leq N + b + 1$$

or, equivalently

$$\sum_{i\in P} x_i(k+1) \leq b + N + 1 + \sum_{i\in P} x_i(0),$$

for all $k \in W$. This completes the proof. ■

Notice that while the RPS stability condition in Theorem 5.4 is more flexible than the condition in Theorem 5.3 (in terms of our ability to design a machine that can achieve stability by speeding it up), the input stream rate constraints are still limited by the maximum machine setup time, regardless of how fast the machine is. This appears to be a fundamental property of RPS policies. Notice also that, if $a_i^{\text{in}'}$ and δ_i' are considered to be inverse rate constraints, the stability condition of Theorem 5.4 can be thought of as reducing to the capacity condition as $s_i \to 0$ for all $i \in P$.

5.2.5 Characterization of the Output Stream

We now consider characterizing the output streams of machines controlled by one of the three policies that we have analyzed. In these characterizations, we will choose the output stream rate constraint to be the same as the input stream rate constraint, so that the machines are *rate preserving*. Of course, the burstiness of the output streams will have to be chosen appropriately. Such a characterization facilitates the interconnection of these machines with each other and with other network elements, which we discuss in the next two sections.

Let the bound from the stability theorems for any of the scheduling policies on every buffer $i \in P$ be B, so that $x_i(k) \leq B$ for all $k \in W$. Choose any times $k_1, k_2 \in W$, such that $k_1 \leq k_2$. Because the machine cannot output parts on output stream $i \in P$ that have not yet arrived at buffer i, we know that

$$\begin{aligned}
\sum_{k=k_1}^{k_2} D_i(k) &\leq x_i(k_1) + \sum_{k=k_1}^{k_2} A_i(k) \\
&\leq B + \sum_{k=k_1}^{k_2} A_i(k) \\
&\leq \operatorname{ceil}\left(a_i^{\text{in}}(k_2 - k_1 + 1) + b_i^{\text{in}}\right) + B\,, \qquad (5.26)
\end{aligned}$$

and we say that machine output stream i has maximum rate a_i^{in} and maximum burstiness $b_i^{\text{in}} + B$.

5.2.6 Network Element Analysis

Because the ultimate goal of this work is the stability analysis of flexible manufacturing systems, we now analyze network elements, other than the isolated machine, which may appear in such systems. Specifically, we analyze the manner in which such network elements transform their input streams into output streams. The four network elements that we consider are: the multiplexer, the demultiplexer, the stream modifier, and the bounded delay.

Multiplexer The type of multiplexer that we consider is a first-come first-serve (FCFS) multiplexer. The FCFS multiplexer can be thought of as a "stream merger," because it merges its N input streams into a single output stream and accumulates no parts at its input. Every part that arrives at the FCFS multiplexer is passed directly to the output stream.

Let $P = \{1, 2, \ldots, N\}$, and let the number of parts to arrive at the multiplexer via input stream $i \in P$ at time $k \in W$ be $A_i(k)$. We require, for all $i \in P$, that $A_i(k)$ satisfy

$$\sum_{k=k_1}^{k_2} A_i(k) \quad \leq \quad \text{ceil}\left(a_i^{\text{in}}(k_2 - k_1 + 1) + b_i^{\text{in}}\right) \tag{5.27}$$

for all $k_1, k_2 \in W$, $k_1 \leq k_2$. Let the number of parts to leave the multiplexer via its sole output stream at time $k \in W$ be $D(k)$.

Because we have assumed that parts arriving at the multiplexer are immediately placed on its output stream, it is clear that

$$\begin{aligned} \sum_{k=k_1}^{k_2} D(k) &= \sum_{i \in P} \sum_{k=k_1}^{k_2} A_i(k) \\ &\leq \sum_{i \in P} \text{ceil}\left(a_i^{\text{in}}(k_2 - k_1 + 1) + b_i^{\text{in}}\right), \end{aligned} \tag{5.28}$$

for all $k_1, k_2 \in W$, $k_1 \leq k_2$. Hence, the output stream of the FCFS multiplexer may be considered to have maximum rate $\sum_{i \in P} a_i^{\text{in}}$ and maximum burstiness $\sum_{i \in P} b_i^{\text{in}}$.

We are not able to distinguish the order of arrival of parts that arrive at the same time $k \in W$. Parts that arrive at the FCFS multiplexer at time k are placed on the output stream at time k in arbitrary order.

Demultiplexer The demultiplexer is a network element that splits one input stream into N output streams. Let $P = \{1, 2, \ldots, N\}$. Let $A(k)$ be the number of parts to arrive at the demultiplexer at time $k \in W$, and let $D(k)$ be the number of parts to leave the demultiplexer at time $k \in W$. For purposes of this analysis, consider parts which arrive at the demultiplexer and are to exit on output stream $i \in P$ to be parts of type i. Let $A_i(k)$ be the number of parts of type $i \in P$ to arrive at the demultiplexer at time $k \in W$, so that

$$\sum_{i \in P} A_i(k) = A(k), \tag{5.29}$$

and

$$D_i(k) = A_i(k) \tag{5.30}$$

for all $k \in W$. Parts of type $i \in P$ arriving on the input stream must satisfy

$$\sum_{k=k_1}^{k_2} A_i(k) \quad \leq \quad \text{ceil}\left(a_i^{\text{in}}(k_2 - k_1 + 1) + b_i^{\text{in}}\right) , \tag{5.31}$$

for all $k_1, k_2 \in W$, $k_1 \leq k_2$. From Equations (5.30) and (5.31), it is clear that

$$\sum_{k=k_1}^{k_2} D_i(k) \quad \leq \quad \text{ceil}\left(a_i^{\text{in}}(k_2 - k_1 + 1) + b_i^{\text{in}}\right) , \tag{5.32}$$

for all $k_1, k_2 \in W$, $k_1 \leq k_2$, and we say that demultiplexer output stream i has maximum rate a_i^{in} and maximum burstiness b_i^{in}.

Stream Modifier The stream modifier is a network element which consists of a buffer and a part flow policy, which selectively queues incoming parts in the buffer or passes them directly through to the output stream. In addition, the policy must decide when to release queued parts into the output stream. The purpose of the stream modifier is to alter the maximum rate and maximum burstiness of its input stream. At time $k \in W$, let the number of parts in the stream modifier buffer be $x(k)$, the number of parts arriving at the stream modifier be $A(k)$, and the number of parts leaving the stream modifier be $D(k)$. In addition, let a^{in}, b^{in}, a^{out} and b^{out} be real, nonnegative constants, which are used to describe the input and output streams of the stream modifier. In order to clarify the following analysis, assume that a^{in} and a^{out} are integer. We specify the behavior of the input and output streams of the stream modifier with respect to constants a^{in}, b^{in}, a^{out}, and b^{out} as follows:

(*i*) For all $k_1, k_2 \in W$, $k_1 \leq k_2$,

$$\sum_{k=k_1}^{k_2} A(k) \leq \text{ceil}\left(a^{\text{in}}(k_2 - k_1 + 1) + b^{\text{in}}\right) , \tag{5.33}$$

and

$$\sum_{k=k_1}^{k_2} D(k) \quad \leq \quad \text{ceil}\left(a^{\text{out}}(k_2 - k_1 + 1) + b^{\text{out}}\right) . \tag{5.34}$$

(*ii*) For all $k' \in W$,

$$\sum_{k=0}^{k'} [A(k) - D(k)] \leq \max\left\{\left(a^{\text{in}} - a^{\text{out}}\right)(k' + 1) + b^{\text{in}} - b^{\text{out}} + 1, \; -x(0)\right\} . \tag{5.35}$$

Note that Equation (5.33) simply specifies the input stream constraint for the stream modifier. Equation (5.34) specifies how we would like the output stream of the stream modifier to behave. Equation (5.35) is included as a constraint on stream modifier behavior to ensure that its buffer is bounded. In fact, because $x(k+1) = x(k) + A(k) - D(k)$ and because

$$\sum_{k=0}^{k'} [A(k) - D(k)] = x(k'+1) - x(0) , \tag{5.36}$$

for all $k' \in W$, we see from Equation (5.35) that

$$x(k'+1) \le \max\left\{x(0) + (a^{\text{in}} - a^{\text{out}})(k'+1) + b^{\text{in}} - b^{\text{out}} + 1, 0\right\} , \tag{5.37}$$

for all $k' \in W$. If we choose $a^{\text{out}} = a^{\text{in}}$, then for all $k \in W$, $x(k) \le x(0) + b^{\text{in}} - b^{\text{out}} + 1$ and the stream modifier buffer is bounded. Notice also that if $a^{\text{out}} = a^{\text{in}}$, then we can choose $b^{\text{out}} = 0$ so that the burstiness is completely removed from the output stream. In this case, if the input stream operates at its maximum rate, then the output stream can operate at its maximum rate and the extra bursts of parts on the input stream, that, by definition, will never total more than b^{in}, will be stored in the stream modifier buffer. Notice that if $a^{\text{out}} < a^{\text{in}}$, then either b^{out} is infinite or we cannot bound $x(k)$ for all $k \in W$. It is also clear that if a^{in} or b^{in} is infinite and a^{out} and b^{out} are finite, no bound exists for $x(k)$ for all $k \in W$.

In the above discussion, we have specified how the stream modifier should work. However, we have yet to specify a policy for the stream modifier which will guarantee that it will perform as specified. Any policy which we specify must guarantee that Equations (5.34) and (5.35) are satisfied.

We now specify a practical policy for the stream modifier, which, we will show, satisfies Equations (5.34) and (5.35) by at every time k releasing the maximum number of parts allowable without violating Equation (5.34). For every time $k \in W$, let $E_k = \{E_k^0, E_k^1, \ldots, E_k^k\}$, where $E_k^{k'}$ is the maximum allowable value of $D(k)$ such that

$$\sum_{l=k'}^{k} D(l) = \text{ceil}\left(a^{\text{out}}(k - k' + 1) + b^{\text{out}}\right) . \tag{5.38}$$

Because the stream modifier cannot violate Equation(5.38) for any k', $k' \le k$, let its policy choose

$$D(k) = \min\left(E_k \cup \{A(k) + x(k)\}\right) . \tag{5.39}$$

Note that the policy in Equation (5.39) is not implementable because as $k \to \infty$, $|E_k| \to \infty$; below we will show how to modify Equation (5.39) so that it is an implementable policy. From the policy in Equation (5.39), we see that the stream modifier will indeed output at every time k the maximum number of parts allowable, up to the number that it has available for output at time k, without violating its output stream constraint.

We now must ask whether our policy will satisfy Equations (5.34) and (5.35). Because our policy in Equation (5.39) guarantees that Equation (5.38) will not be violated for any $k', k \in W$, $k' \leq k$, it is clear that it satisfies Equation (5.34), because Equation (5.38) essentially amounts to a different characterization of Equation (5.34).

We now concern ourselves with satisfying Equation (5.35). Choose any time $k' \in W$. If

$$D(k') \quad = \quad \min\{E_{k'} \cup \{A(k') + x(k')\}\} \quad = \quad A(k') + x(k')\;, \qquad (5.40)$$

then because $x(k'+1) = 0$ (the stream modifier buffer is cleared by the policy's choice of $D(k')$), it is clearly the case that

$$\sum_{k=0}^{k'} (A(k) - D(k)) \quad = \quad -x(0), \qquad (5.41)$$

so that Equation (5.35) holds. If, on the other hand,

$$D(k') \quad = \quad \min\{E_{k'} \cup \{A(k') + x(k')\}\} \quad = \quad \min(E_{k'}), \qquad (5.42)$$

then from the definitions of E_k' in Equation (5.38) and the stream modifier policy in Equation (5.39), there is some $k_1 \in W$, $k_1 \leq k'$, such that

$$\sum_{k=k_1}^{k'} D(k) \quad = \quad \mathrm{ceil}\left(a^{\mathrm{out}}(k' - k_1 + 1) + b^{\mathrm{out}}\right)\;. \qquad (5.43)$$

Below we define a recursive procedure whose goal is to define a time $k_* \in W$ which we use later in the analysis. Upon initially entering the procedure, let $i = 0$.

1. If $i = 0$, let $n_i = k'$; otherwise, let $n_i = m_{i-1} - 1$.

2. If $D(n_i) = \min(E_{n_i})$, then find the smallest $q \in W$, such that

$$\sum_{k=q}^{n_i} D(k) \quad = \quad \mathrm{ceil}\left(a^{\mathrm{out}}(n_i - q + 1) + b^{\mathrm{out}}\right)\;, \qquad (5.44)$$

   and let $m_i = q$.

3. If $m_i = 0$, then let $k_* = 0$ and stop.

4. If $D(m_i - 1) < \min(E_{m_i - 1})$, then let $k_* = m_i - 1$ and stop; otherwise, let $i = i + 1$ and return to step 1.

Notice that the above procedure will always terminate. If $k_* = 0$, then the entire range of times $[0, k']$ is composed of adjacent subranges of times $[m_i, n_i]$, $i = 0, 1, 2, \ldots, Q$, where $n_0 = k'$, $m_Q = 0$, $n_i = m_{i-1} - 1$ for all $i = 1, 2, \ldots, Q$, and

$$\sum_{k=m_i}^{n_i} D(k) = \text{ceil}\left(a^{\text{out}}(n_i - m_i + 1) + b^{\text{out}}\right), \tag{5.45}$$

for all $i = 0, 1, 2, \ldots, Q$. Hence, we see that

$$\sum_{k=0}^{k'} D(k) = \sum_{i=0}^{Q} \sum_{m_i}^{n_i} \text{ceil}\left(a^{\text{out}}(n_i - m_i + 1) + b^{\text{out}}\right) \tag{5.46}$$

and (because $\text{ceil}(a) + \text{ceil}(b) \geq \text{ceil}(a + b)$) that

$$\sum_{k=0}^{k'} D(k) \geq \text{ceil}\left(a^{\text{out}}(k' + 1) + (Q + 1)b^{\text{out}}\right). \tag{5.47}$$

Because we have established that our policy obeys Equation (5.34), we see that either Q or b^{out} must equal 0, so that

$$\sum_{k=0}^{k'} D(k) = \text{ceil}\left(a^{\text{out}}(k' + 1) + b^{\text{out}}\right). \tag{5.48}$$

Therefore,

$$\begin{aligned}\sum_{k=0}^{k'} (A(k) - D(k)) &\leq \text{ceil}\left(a^{\text{in}}(k' + 1) + b^{\text{in}}\right) - \\ &\quad\ \text{ceil}\left(a^{\text{out}}(k' + 1) + b^{\text{out}}\right) & (5.49) \\ &\leq \left(a^{\text{in}} - a^{\text{out}}\right)(k' + 1) + b^{\text{in}} - b^{\text{out}} + 1, & (5.50)\end{aligned}$$

so that for $D(k') = \min(E_{k'})$ and $k_* = 0$, our policy satisfies Equation (5.35).

If $k_* > 0$, then the entire range of times $[k_* + 1, k']$ is composed of adjoining subranges of times $[m_i, n_i]$, $i = 0, 1, 2, \ldots, Q$, where $n_0 = k'$, $m_Q = k_* + 1$, $n_i = m_{i-1} - 1$ for all $i = 1, 2, \ldots, Q$, and

$$\sum_{k=m_i}^{n_i} D(k) = \text{ceil}\left(a^{\text{out}}(n_i - m_i + 1) + b^{\text{out}}\right), \tag{5.51}$$

for all $i = 0, 1, 2, \ldots, Q$. Similar to before, it follows that

$$\sum_{k=k_*+1}^{k'} D(k) = \text{ceil}\left(a^{\text{out}}(k' - k_*) + b^{\text{out}}\right). \tag{5.52}$$

Table 5.1 E_k

$E_k^k = \text{ceil}\left(a^{\text{out}} + b^{\text{out}}\right)$
$E_k^{k-1} = \text{ceil}\left(2a^{\text{out}} + b^{\text{out}}\right) - D(k-1)$
$\vdots$
$E_k^0 = \text{ceil}\left(a^{\text{out}}(k+1) + b^{\text{out}}\right) - \sum_{n=0}^{k-1} D(n)$

Because $\sum_{k=0}^{k_*} (A(k) - D(k)) \le 0$, we see that

$$\begin{aligned}\sum_{k=0}^{k'} (A(k) - D(k)) &= \sum_{k=0}^{k_*} (A(k) - D(k)) \\ &\quad + \sum_{k=k_*+1}^{k'} (A(k) - D(k)) &(5.53)\\ &\le \sum_{k=k_*+1}^{k'} (A(k) - D(k)) &(5.54)\\ &\le \text{ceil}\left(a^{\text{in}}(k'-k_*) + b^{\text{in}}\right) - \\ &\quad \text{ceil}\left(a^{\text{out}}(k'-k_*) + b^{\text{out}}\right) &(5.55)\\ &\le \left(a^{\text{in}} - a^{\text{out}}\right)(k'-k_*) + b^{\text{in}} - b^{\text{out}} + 1 &(5.56)\\ &\le \left(a^{\text{in}} - a^{\text{out}}\right)(k'+1) + b^{\text{in}} - b^{\text{out}} + 1\,, &(5.57)\end{aligned}$$

so that for $D(k') = \min(E_{k'})$ and $k_* > 0$, our policy satisfies Equation (5.35). Hence, we have shown that Equations (5.34) and (5.35) are satisfied by our stream modifier policy.

We have shown that the above policy for selecting $D(k)$ satisfies conditions (i) and (ii), we now show how to recursively calculate E_k for all $k \in W$. From Equation (5.38), we can form Tables 5.1 and 5.2.

From the form of the expressions in Tables 5.1 and 5.2, notice that

$$E_{k+1}^{k'} = E_k^{k'} + a^{\text{out}} - D(k) \tag{5.58}$$

for all k', $k' \le k$, so that

$$\min(E_{k+1} - \{E_{k+1}^{k+1}\}) = \min(E_k) + a^{\text{out}} - D(k)\,. \tag{5.59}$$

It is clear, then, that if we define the set

$$\bar{E}_k = \left\{\min(\bar{E}_{k-1}) + a^{\text{out}} - D(k-1)\right\} \cup \left\{\text{ceil}\left(a^{\text{out}} + b^{\text{out}}\right)\right\},$$

Table 5.2 E_{k+1}

$E_{k+1}^{k+1} = \text{ceil}\left(a^{\text{out}} + b^{\text{out}}\right)$
$E_{k+1}^{k} = \text{ceil}\left(2a^{\text{out}} + b^{\text{out}}\right) - D(k)$
$E_{k+1}^{k-1} = \text{ceil}\left(3a^{\text{out}} + b^{\text{out}}\right) - D(k-1) - D(k)$
$\vdots$
$E_{k+1}^{0} = \text{ceil}\left(a^{\text{out}}(k+2) + b^{\text{out}}\right) - \sum_{n=0}^{k} D(n)$

for all $k \in W$, $k \neq 0$ and if we let

$$\bar{E}_0 = \left\{\text{ceil}\left(a^{\text{out}} + b^{\text{out}}\right)\right\} , \tag{5.60}$$

then

$$\min(E_k) = \min\left(\bar{E}_k\right) . \tag{5.61}$$

Hence, we can define our policy for all $k \in W$ as

$$D(k) = \min\left\{\bar{E}_k \cup \{A(k) + x(k)\}\right\} . \tag{5.62}$$

Our stream modifier is similar to the (σ, ρ) regulator in [23]. Our stream modifier allows slightly more flexibility, however, because the (σ, ρ) regulator always preserves rate from its input stream to its output stream. Although our stream modifier has the ability to alter rate as well as burstiness, if the output stream rate constraint is to be less than the input stream rate constraint, either the output stream burstiness constraint or the steam modifier buffer must be infinite.

Bounded Delay The bounded delay is a network element which may delay parts by no more than a given number of periods, T. At time $k \in Z$, let the number of parts entering the delay element be $A(k)$ and the number of parts leaving the delay element be $D(k)$. In addition, let a^{in} and b^{in} be nonnegative, real constants. Parts arrive at the delay element according to

$$\sum_{k=k_1}^{k_2} A(k) \leq \text{ceil}\left(a^{\text{in}}(k_2 - k_1 + 1) + b^{\text{in}}\right) , \tag{5.63}$$

for all $k_1, k_2 \in W$, $k_2 \geq k_1$. Parts exiting the delay element must obey

$$\sum_{k=k_1}^{k_2} D(k) \leq \text{ceil}\left(a^{\text{in}}(k_2 - k_1 + T + 1) + b^{\text{in}}\right) , \tag{5.64}$$

for all $k_1, k_2 \in W$, $k_1 \leq k_2$. Following [23], this result follows directly from Equation (5.63) and the maximum delay, T:

$$\sum_{k=k_1}^{k_2} D(k) \leq \sum_{k=k_1-T}^{k_2} A(k) \quad \text{for all } k_1, k_2 \in Z,\ k_2 \geq k_1 \tag{5.65}$$

$$\leq \text{ceil}\left(a^{\text{in}}(k_2 - k_1 + T + 1) + b^{\text{in}}\right) . \tag{5.66}$$

Hence, we see that the bounded delay element preserves rate and increases burstiness.

5.3 STABILITY OF DISTRIBUTED FMS

In this section we consider a general FMS, composed of the network elements of the previous section and the single machine, controlled by any of the policies that we have analyzed (CAF, CAOB, RPS). We first establish two sets of conditions for which such an FMS is stable. The second set of conditions involves the augmentation of the network with stream modifiers. We next outline a procedure for the computation of buffer bounds in a stable network. Finally, we discuss a virtual implementation of some of the stream modifiers in the network.

5.3.1 Modeling the FMS

We represent the FMS by $(G, A_i)_{i=1}^{N}$, where N is now the number of paths on which parts may flow through the network, G is the set of network elements (machine and stream modifier buffers, multiplexers, demultiplexers, bounded delays) and each $A_i \subset G \times G$ is the set of directed arcs representing the connections between elements on the *ith* path. We define G precisely by

$$G = MB^{\text{f}} \cup MB^{\text{r}} \cup MB^{\text{b}} \cup SM \cup FM \cup DM \cup BD \cup EP \cup DP , \tag{5.67}$$

where MB^{f} is the set of CAF machine buffers, MB^{r} is the set of RPS machine buffers, MB^{b} is the set of CAOB machine buffers, SM is the set of stream modifiers, FM is the set of FCFS multiplexers, DM is the set of demultiplexers, BD is the set of bounded delays, EP is the set of "entry points," and DP is the set of "departure points." Let $P = \{1, 2, \ldots, N\}$. There are N entry points and each one represents the place at which parts of type $i \in P$ first enter the FMS. Similarly, there are N departure points, with each one representing the place at which parts of type $i \in P$ depart the network. Notice that a single machine is, in general, represented by more than one element.

Define the set A, such that

$$A = \bigcup_{i=1}^{N} A_i , \quad \text{where } A_i \subset G \times G . \tag{5.68}$$

Each A_i contains a sequence of arcs which represent a single path through the network and N is the number of such paths. Every part that enters the network on path $i \in P$ must remain on path i until it exits the network, and we will refer to parts which traverse the network on path $i \in P$ as parts of type i. For example, if $A_1 = \{(g_1, g_3), (g_3, g_5), (g_5, g_9)\}$, $g_i \in G$, then a part of type 1 traversing path 1 begins at element g_1 (entry point), moves to element g_3, moves to element g_5, and exits the network at element g_9 (departure point). Let L_i be the number of distinct $g_i \in G$ elements on path $i \in P$. We place the following restrictions on A_i, $i \in P$:

(*i*) For all $i \in P$, if $A_i = \{(\alpha_j, \beta_j) : \alpha_j, \beta_j \in G, \ j = 1, 2, \ldots, L_i - 1\}$, then

(*a*) $\alpha_1 \in EP$, $\beta_{L_i - 1} \in DP$, $\alpha_j \notin EP \cup DP$ for all $j \neq 1$, and $\beta_m \notin DP \cup EP$ for all $m \neq L_i - 1$ (i.e., the path begins at an entry point and ends at a departure point and there are no other entry points or departure points in the path).

(*b*) For every $j \in \{1, 2, \ldots, L_i - 2\}$, $\beta_j = \alpha_{j+1}$ (i.e., the path is continuous).

(*c*) For every $j, m \in \{1, 2, \ldots, L_i - 1\}$, $\alpha_j \neq \alpha_m$ if $j \neq m$ (i.e., parts do not revisit the same element).

(*ii*) For all $\eta \in DM$, if $Q_\eta = \{i \in P : (\alpha, \eta) \in A_i, \ \alpha \in G\}$ and $R_\eta = \{(\eta, \beta) \in A : \beta \in G\}$, then $|Q_\eta| \geq |R_\eta|$ (i.e., a demultiplexer cannot "split" a single path).

It is interesting to note that because of the presence of multiplexers in the network, it is possible that a single arc which is contained in A may be contained in more than a single A_i, $i \in P$. For instance, it is possible that $\gamma_1 \in A$, $\gamma_1 \in A_1$, and $\gamma_1 \in A_4$. In this case, there is one physical link, γ_1, between two network elements, and this link carries parts of parts of type 1 and parts of type 4 (i.e., paths 1 and 4 flow over the same physical link). Also note that while we do not allow a part to revisit the same element, we do allow parts to revisit the same machine.

For every arc, $\gamma = (\alpha, \beta)$, $\gamma \in A$, in the network, let $M_\gamma(k)$ be the number of parts to leave element α and arrive at element β at time $k \in W$. Notice that parts traverse arcs instantaneously (network delays may be modeled with the bounded delay element). Because it may be convenient to refer to an arc by the element which parts arrive at or depart from via the arc, we give two further definitions. If $\gamma \in A$, $\gamma = (\alpha, \beta)$, and $\alpha \notin DM$, define $M_\alpha^{\text{out}}(k) = M_\gamma(k)$ (the $\alpha \notin DM$ restriction is necessary because there is more than one arc on which parts exit a demultiplexer). If $\gamma \in A$, $\gamma = (\alpha, \beta)$, and $\beta \notin FM$, define $M_\beta^{\text{in}}(k) = M_\gamma(k)$ (the $\beta \notin FM$ restriction is necessary, because there is more than one arc on which parts arrive at a multiplexer).

Let $MB = MB^{\text{f}} \cup MB^{\text{r}} \cup MB^{\text{b}}$, and let N^{f} denote the number of CAF machines in the network, N^{r} denote the number of RPS machines in the network, and N^{b} denote the number of CAOB machines in the network. Let

$MB^{\mathrm{f}} = \bigcup_{i=1}^{N^{\mathrm{f}}} MB_i^{\mathrm{f}}$, where MB_i^{f} is the set of all buffers of CAF machine i. Let $MB^{\mathrm{r}} = \bigcup_{i=1}^{N^{\mathrm{r}}} MB_i^{\mathrm{r}}$, where MB_i^{r} is the set of all buffers of RPS machine i. Let $MB^{\mathrm{b}} = \bigcup_{i=1}^{N^{\mathrm{b}}} MB_i^{\mathrm{b}}$, where MB_i^{b} is the set of all buffers of CAOB machine i. For every $\eta \in MB \cup SM$, let $x_\eta(k)$ be the number of parts in buffer η at time $k \in W$. For every $\eta \in MB$ let δ_η and d_η be the minimum and maximum processing rates, respectively, of machine buffer η (see Equation (5.2)), let N_η be the number of buffers in the machine to which η belongs, and let $\bar{s}_\eta$ be the maximum set-up time of the machine to which η belongs.

5.3.2 Designing Stable Distributed FMS

Notice that under the scheduling policies CAF, RPS, and CAOB, the isolated machine is rate preserving. In addition, the multiplexer, demultiplexer, bounded delay, and stream modifiers are rate preserving. While the machine and bounded delay can introduce burstiness into part streams, the stream modifier can be designed to remove it. In this section, we are concerned with establishing conditions under which we are guaranteed that an FMS which is composed of a set of interconnected elements is stable. First we generate a set of "network compatibility constraints" which establish a set of constraints on interconnections of stable machines that guarantee that the FMS will be stable. Following this, we show how to *design* stream modifiers for an FMS so that (a) it will be stable and (b) we can calculate the buffer bounds. The key feature that allows the stream modifier to stabilize an FMS is its ability to remove burstiness on part streams and thereby allow the network compatibility constraints to be met. (Note: In order to avoid the notational complexity of appropriately defining "inverse rate" constraints, we will only consider the RPS stability conditions of Theorem 5.3. However, extending the following results to include machines governed by the RPS stability conditions of Theorem 5.4 is easily done.)

We say that *network compatibility constraints* are satisfied if for every $\eta \in MB \cup SM$ there exist constants $a_\eta^{\mathrm{in}} < \infty$, $b_\eta^{\mathrm{in}} < \infty$, $a_\eta^{\mathrm{out}} < \infty$, and $b_\eta^{\mathrm{out}} < \infty$, such that for all $k_1, k_2 \in W$, $k_1 \leq k_2$,

$$\sum_{k=k_1}^{k_2} M_\eta^{\mathrm{in}}(k) \quad \leq \quad \mathrm{ceil}\left(a_\eta^{\mathrm{in}}(k_2 - k_1 + 1) + b_\eta^{\mathrm{in}}\right) \tag{5.69}$$

and

$$\sum_{k=k_1}^{k_2} M_\eta^{\mathrm{out}}(k) \quad \leq \quad \mathrm{ceil}\left(a_\eta^{\mathrm{out}}(k_2 - k_1 + 1) + b_\eta^{\mathrm{out}}\right), \tag{5.70}$$

where

(*i*) for all $j \in \{1, 2, \ldots, N^{\mathrm{f}}\}$,

$$\sum_{\eta \in MB_j^{\mathrm{f}}} \frac{a_\eta^{\mathrm{in}}}{\delta_\eta} < 1 , \tag{5.71}$$

(*ii*) for all $\eta \in MB_j^{\mathrm{r}}$, $j \in \{1, 2, \ldots, N^{\mathrm{r}}\}$,

$$a_\eta^{\mathrm{in}} \leq \frac{1}{N_\eta(\bar{s}_\eta + 1)} , \tag{5.72}$$

(*iii*) for all $\eta \in MB_j^{\mathrm{b}}$, $j \in \{1, 2, \ldots, N^{\mathrm{b}}\}$,

$$\frac{a_\eta^{\mathrm{in}}}{\delta_\eta} < \frac{1}{N_\eta} , \tag{5.73}$$

and

(*iv*) for all $\eta \in SM$, $a_\eta^{\mathrm{out}} \geq a_\eta^{\mathrm{in}}$.

Notice that Equations (5.69) and (5.70) specify stream constraints for the FMS that via (*i*)–(*iv*) guarantee that all the elements in the FMS are stable. Hence, we obtain the following result:

Theorem 5.5 (FMS Stability: Compatibility Constraints): *The FMS* $(G, A)_{i=1}^{N}$ *is stable if the network compatibility constraints are satisfied.*

Proof: The result is a direct consequence of the stability theorems for the local scheduling policies and the fact that if its output stream rate constraint is at least as large as its input stream rate constraint, then the number of parts in the buffer of a stream modifier can grow no larger than the number there initially plus the maximum burstiness of its input stream plus one. ■

For a given FMS, it may be difficult or impossible to verify the conditions of Theorem 5.5, and, from [43] and [38], we know that a very real potential for instability exists, particularly in systems that include machines involved in part loops (i.e., machine *a* and machine *b* are in a part loop if machine *a* feeds machine *b* and machine *b* feeds machine *a*). A network with these types of part loops is called *nonacyclic* in [43].

We now show how to employ stream modifiers to not only stabilize general FMS, but to allow the computation of buffer bounds. Although this section of the work is similar to the regulator stabilization technique of [38], it was developed independently and allows for the calculation of buffer bounds in general nonacyclic FMS, including nonacyclic FMS containing machines governed by the RPS policy.

Consider an FMS $(G', A'_i)_{i=1}^N$, which is identical to $(G, A_i)_{i=1}^N$ except that FMS $(G', A'_i)_{i=1}^N$ has a stream modifier directly preceding every machine buffer so that for all $\eta \in MB$, there is some $\gamma \in A'$ such that $\gamma = (\eta', \eta)$ for some $\eta' \in SM$. As in the definition of A, let $A' = \bigcup_i A'_i$, where each A'_i contains a sequence of arcs which represent a single path through the network.

For all $\eta \in EP$, if $(\eta, \beta) \in A_i$ for some $\beta \in G'$, choose constants a^i and b^i, such that

$$\sum_{k=k_1}^{k_2} M_\eta^{\text{out}}(k) \leq \text{ceil}\left(a^i(k_2 - k_2 + 1) + b^i\right) , \tag{5.74}$$

for all $k_1, k_2 \in W$, $k_1 \leq k_2$. Clearly, a^i and b^i are the maximum rate and burstiness, respectively, of network input stream i (i.e., the stream constraints for part type $i \in P$ when it first enters the FMS). For every $\gamma \in A'$, define the set P_γ so that $P_\gamma = \{i \in P : \gamma \in A'_i\}$ (set of all part types that flow over one physical link), and define the constant a^γ, so that

$$a^\gamma = \sum_{i \in P_\gamma} a^i . \tag{5.75}$$

Notice that, even though $\gamma \in A'$, may lie "deep in the FMS," the constant a^γ is defined in terms of the rate constraints of the parts of type $i \in P_\gamma$ as they first enter the network. For notational ease in the following analysis, if $\gamma \in A'$ and $\gamma = (\alpha, \eta)$ for some $\alpha \in G'$, then define $a^\eta \triangleq a^\gamma$.

In the theorem below, we establish conditions on the network input rate constraints and on the stream modifier policies to guarantee stability of general nonacyclic FMS.

Theorem 5.6 (FMS Stability): *The FMS $(G', A'_i)_{i=1}^N$ is stable if:*

(i) For all $j \in \{1, 2, \ldots, N^f\}$,

$$\sum_{\eta \in MB_j^f} \frac{a^\eta}{\delta_\eta} < 1 . \tag{5.76}$$

(ii) For all $\eta \in MB_j^r$, $j \in \{1, 2, \ldots, N^r\}$,

$$a^\eta \leq \frac{1}{N_\eta(\bar{s}_\eta + 1)} . \tag{5.77}$$

(iii) For all $\eta \in MB_j^b$, $j \in \{1, 2, \ldots, N^b\}$,

$$\frac{a^\eta}{\delta_\eta} < \frac{1}{N} . \tag{5.78}$$

(*iv*) *For all* $\gamma \in \{(\alpha, \beta) \in A' : \alpha \in SM,\ \beta \in G'\}$, *and for some* $b' < \infty$,

$$\sum_{k=k_1}^{k_2} M_\gamma(k) \leq ceil\left(a^\gamma(k_2 - k_1 + 1) + b'\right) , \qquad (5.79)$$

for all $k_1, k_2 \in W,\ k_1 \leq k_2$.

Proof: We begin by listing the following facts:

(*i*) By Theorem 5.5 and the assumptions of the theorem to be proved, it is clear that for any $\eta \in MB$, $x_\eta(k) \leq B_\eta$ for some $B_\eta < \infty$ and for all $k \in W$. From the machine output stream characterization in Equation (5.26), we see that for any $\eta \in MB$,

$$\sum_{k=k_1}^{k_2} M_\gamma(k) \leq \text{ceil}\left(a^\gamma(k_2 - k_1 + 1) + b'\right), \quad \gamma = (\eta, \beta) \in A',\ \beta \in G, \qquad (5.80)$$

for all $k_1, k_2 \in W$, $k_1 \leq k_2$ and for some $b' < \infty$.

(*ii*) From the network element output stream characterizations in Equations (5.28) (multiplexer), (5.32) (demultiplexer), and (5.64) (bounded delay), it is evident for any $\eta \in FM \cup DM \cup BD$ that if, for all $\gamma \in \{(\alpha, \eta) \in A' : \alpha \in G'\}$,

$$\sum_{k=k_1}^{k_2} M_\gamma(k) \leq \text{ceil}\left(a^\gamma(k_2 - k_1 + 1) + b'\right) , \qquad (5.81)$$

for all $k_1, k_2 \in W$, $k_1 \leq k_2$, and for some $b' < \infty$, then for all $\gamma' \in \{(\eta, \beta) \in A' : \beta \in G'\}$,

$$\sum_{k=k_1}^{k_2} M_{\gamma'}(k) \leq \text{ceil}\left(a^{\gamma'}(k_2 - k_1 + 1) + b''\right) , \qquad (5.82)$$

for all $k_1, k_2 \in W$, $k_1 \leq k_2$, and for some $b'' < \infty$.

Given facts (*i*) and (*ii*) above, along with assumption (*iv*) in the statement of the theorem, it is clear that for all $\gamma \in A'$,

$$\sum_{k=k_1}^{k_2} M_\gamma(k) \leq \text{ceil}\left(a^\gamma(k_2 - k_1 + 1) + b'\right) , \qquad (5.83)$$

for all $k_1, k_2 \in W$, $k_1 \leq k_2$ and for some $b' < \infty$. From this fact and Equation (5.37), we see that for every $\eta \in SM$, there exists a bound B_η such that $x_\eta(k) \leq B_\eta$ for all $k \in W$. Hence, for every $\eta \in MB \cup SM$, there exists a bound B_η, such that $x_\eta(k) \leq B_\eta$ for all $k \in W$, and the FMS is stable. ∎

Notice that, for a given network, it is very easy to test conditions (i), (ii), and (iii) of Theorem 5.6. If the condition is not met for a given set of maximum network input rates, it is clear which network input streams need to be affected so that the condition is satisfied. Condition (iv) of Theorem 5.6 simply specifies, to a certain extent, the part flow policy for the stream modifiers in the network. From condition (iv), we see that the output stream rate constraint of every stream modifier in the network is fixed, and as it turns out, the output stream rate constraints are fixed to match the input stream rate constraints.

5.3.3 Computing the Buffer Bounds

Once we have satisfied the conditions for the stability of an FMS (G', A'), so that for all $\eta \in MB \cup SM$, $x_\eta(k) \leq B_\eta$ for some $B_\eta < \infty$ and for all $k \in W$, it is straightforward to compute such bounds $B_\eta < \infty$ for all $\eta \in MB \cup SM$.

From the proof of Theorem 5.6, we see that the maximum rate constraint is a^γ and the maximum burstiness constraint is finite for all parts traveling on any arc $\gamma \in A'$, so that

$$\sum_{k=k_1}^{k_2} M_\gamma(k) \leq \text{ceil}\left(a^\gamma(k_2 - k_1 + 1) + b^\gamma\right) \qquad (5.84)$$

for some $b^\gamma < \infty$ and for all $k_1, k_2 \in W$, $k_1 \leq k_2$. In addition, we know that for all

$$\gamma \in \{(\eta, \beta) \in A'_i : \eta \in EP, \beta \in G'\},$$

$b^\gamma = b^i$, where $\gamma \in A'_i$. Because we know from Equations (5.26) (single machine), (5.28) (multiplexer), (5.32) (demultiplexer), (5.34) (stream modifier), and (5.64) (bounded delay) how each network element transforms its maximum input stream rate and burstiness constraints into maximum output stream rate and burstiness constraints, we can simply trace each path from entry point to departure point, calculating maximum burstiness constraints for every arc. When we come to a stream modifier, we specify its maximum output stream burstiness constraint (if the stream modifier feeds a machine buffer, we are also specifying the maximum input stream burstiness constraint of the machine buffer).

From Equations (5.37) and (5.34), it is clear that by specifying the maximum burstiness constraint of the output stream of a stream modifier $\eta \in SM$, we also specify the bound on its buffer, B_η. Once the maximum burstiness constraint has been specified for the input stream of every $\eta \in MB$, the buffer bounds, B_η, follow directly from the proofs of stability for the isolated machine scheduling policies.

5.3.4 Artificial Implementation of the Stream Modifiers

The stream modifiers preceding the machine buffers may all be removed, without affecting performance, if the machines change the manner in which they view their buffers. Each machine buffer now must consist of two segments: the virtual machine buffer and the virtual stream modifier buffer. Each machine buffer is, of course, just one buffer physically, but the distinction between its two segments is important. Only parts that would have arrived at the machine buffer had its feeding stream modifier remained in place are considered to be in the virtual machine buffer. The remainder of parts in the actual machine buffer are considered to be in the virtual stream modifier buffer. All machine part servicing policy decisions are based only on the contents of the virtual machine buffer. The policy has no knowledge of the contents of the virtual stream modifier buffer. In this manner, we can eliminate all of the stream modifiers that we added to G to form G'. Of course, the true bound on any actual machine buffer is now the sum of the bounds of its virtual machine buffer and its virtual stream modifier buffer.

Notice that implementing the stream modifiers artificially as above may result in the truncation of clearing policy production runs before the actual machine buffer is empty. This behavior is reminiscent of the behavior of the Universally Stabilizing Supervisory Mechanism (USSM) of [43]. The difference between the action of the USSM and the machines in the network described above is that the USSM truncates production runs once they have continued for a particular length of time and the machines in the network described above truncate production runs when the virtual machine buffer being serviced is empty. As a consequence of this difference in action, we are able to identify specific bounds for machine buffers, while in a network under the control of USSM, it is only known that the buffers are bounded.

5.3.5 Buffer Bounds: Theoretical and Simulated

In this section, we employ simulation to get some idea of how sharp our theoretical buffer bounds are. We calculate the buffer bounds according to the methodology outlined in the previous section. We simulate the FMS of Figure 5.2. The rate and burstiness parameters of the sole network input stream and the processing rates for each buffer are apparent from the figure. The stream modifiers are not explicitly shown as they are implemented artificially. We assume that there are no transportation delays in the network. We use the CLB policy to control each machine. A summary of the theoretical and simulated buffer bounds, is given in Table 5.3.

Note that each numerical entry in Table 5.3 corresponds to a bound on the sum of all buffers of a particular machine. In this example, the theoretical bounds are clearly not very tight. While we have just given one example where the theoretical bounds are conservative, this, unfortunately, will be the case most often. This is primarily due to the fact that the buffer bounds

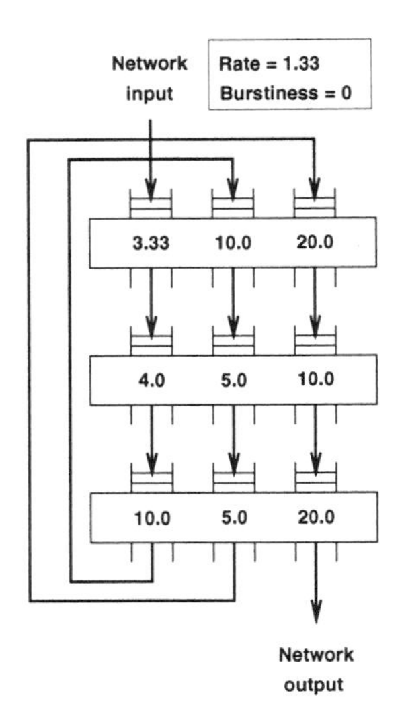

Fig. 5.2 Example FMS.

Table 5.3 Sum of buffer bounds, theoretical and simulated.

	Theoretical	Simulated
Machine 1	2264	1627
Machine 2	4784	2144
Machine 3	3194	1972

of the machine scheduler theorems cannot be specified for individual buffers but only for the sum of buffers. This type of result is not without precedent, however. It is common in system analysis to resort to norm-based results when more exact results are unattainable.

5.4 FMS SCHEDULING: A CENTRALIZED APPROACH

The primary advantage of an FMS composed of individual machines, each with its own scheduling policy that utilizes only local information, is that the individual machines need not communicate with one another so that real-time implementation is simplified. However, for many modern FMS, it is quite realistic to allow intermachine communications. For the remainder of this chapter we seek to exploit this fact by developing scheduling policies that incorporate information from other parts of the network that can be useful in making efficient scheduling decisions. In using more "global" information, we are careful to minimize the level of necessary communications so that our global policies are implementable in real-time, just as the local policies of the previous sections.

In the next section, we define and analyze a class of global scheduling policies that we call Global Synchronous Clearing (GSC) policies. As is evident from the name, policies in this class exhibit two properties: they are clearing policies and they exploit machine synchronization. Similar to the way in which local policies select a buffer to service from among the buffers of a single machine, GSC policies select from among a set of *paths* to service. A path is a set of topologically consecutive buffers which can be serviced simultaneously. In general, a GSC policy will choose from among *sets* of paths to process. When a GSC policy chooses a set of paths to process, all buffers in each path in the set are processed simultaneously (hence, all paths in a set must be able to be processed at the same time). Once a GSC policy begins servicing a set of paths, servicing continues until all paths in the set are clear of parts.

Here, we will analyze two GSC policies: the global synchronous clear-a-fraction (GSCAF) policy and the global synchronous periodic clearing (GSPC) policy. The GSCAF policy is a global extension of the local CAF policy that was studied earlier in this chapter in that it determines which set of paths to service by summing up all parts in each set of paths and determining which sets have sums of parts greater than a fixed fraction of the total number of parts in the FMS. We provide sufficient conditions for the stable operation (all buffers remain bounded for all time) of FMS under GSCAF control. In addition, our analysis provides buffer bounds for stable FMS. The GSPC policy simply enforces a periodic service sequence. We provide sufficient conditions for the stable operation of FMS under GSPC control. Again, our analysis provides buffer bounds for stable FMS. In fact, for an important class of FMS, our GSPC analysis yields a very precise characterization of FMS behavior.

Intuitively, it seems that if we allow a scheduling policy to have access to global information, it ought to be able to yield performance superior to that of a network of isolated schedulers. For the class of global scheduling policies that we propose, we illustrate, via a battery of comparative simulations, that it is often the case that a global scheduler is able to outperform a network of isolated schedulers (i.e., a decentralized scheduler). However, we show that this is not always the case and that the relative degree of success of the global scheduler with respect to a network of isolated schedulers is dependent upon network topology and machine parameters. In so doing, we impart the reader with some heuristic guidelines for determining the applicability of global scheduling policies to particular FMS.

5.4.1 FMS Description and Notation

Let there be N "paths" within the FMS. The three important attributes of any path are (1) all of its buffers may be serviced at the same time, (2) its buffers are directly connected in the network, and (3) that if a buffer is on one path, it is not on another path and all buffers are on a path. For example,

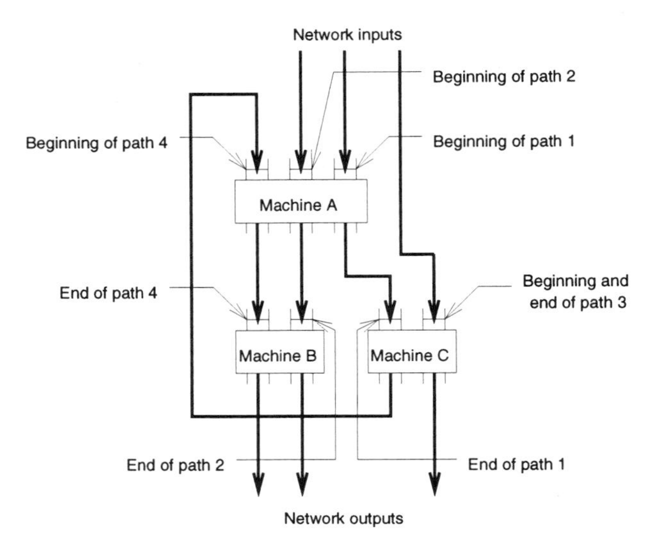

Fig. 5.3 Example FMS with paths labeled.

in the network of Figure 5.3 we can define four paths which begin and end as indicated. Notice that the beginning and end of a path are defined as the first and last buffer in the path, respectively. Let each buffer be referred to by a coordinate (i, j), where i is the path number and j is the buffer number along the path. For example, in the FMS of Figure 5.3 the buffer that ends path 4 is referred to by coordinate $(4, 2)$. Let Q be the set of paths which originate at network inputs and let R be the set of paths which originate at network re-entry points. For the FMS of Figure 5.3, $Q = \{1, 2, 3\}$ and $R = \{4\}$.

Let the level of buffer (i, j) at time t be $x_{i,j}(t)$, and let $X(t) = \{x_{i,j}(t)\}$ be the set of all of the buffer levels at time t. Let a_i be the part arrival rate (in parts per unit time) at network input buffer $(i, 1)$, $i \in Q$. For convenience, if $i \in R$, we let $a_i = 0$. Let s_i be the maximum set-up time of all the buffers on path i. For example, if for path 1 it takes two time units for machine A to configure itself to process parts from buffer $(1, 1)$ and it takes one time unit for machine C to configure itself to process parts from buffer $(1, 2)$, then $s_1 = 2$. Let τ_i be the minimum processing rate for buffers on path i. All buffers in path i are serviced at rate τ_i and the sum of the transport delays between buffers on path i (including the transport delay between the last buffer in path i and the head buffer of a downstream path, if applicable) is D_i. Let P_i be the number of paths, including i, that a part in buffer $(i, 1)$ will traverse before exiting the FMS. For example, in the FMS of Figure 5.3, $P_1 = 2$ and $P_i = 1$ for $i = 2, 3, 4$. If $P_i > 1$, define H_i to be the path that is fed by path i. For example, in the FMS of Figure 5.3, $H_1 = 4$.

Choose a set of paths such that all paths in the set can be serviced simultaneously. Let $S = \{S_i : 1 \leq i \leq M\}$ be the set of such sets of paths. We require that the set S satisfy the following condition: for all paths $i \in \{1, 2, \ldots, N\}$, there is some $j \in \{1, 2, \ldots, M\}$ such that $i \in S_j$ (each path is in at least one S_j). For example, in the FMS of Figure 5.3, we can define S as $S = \{\{1\}, \{2, 3\}, \{3, 4\}\}$, where $S_1 = \{1\}$, $S_2 = \{2, 3, \}$, and $S_3 = \{3, 4\}$.

The production scheduling policies for the FMS that we now consider are what we will call "Global Synchronous Clearing (GSC) policies." A GSC policy is one in which a production run (i.e., a time period during which parts from a particular set of buffers are processed) is begun by choosing a set of paths $S_i \in S$. All buffers on paths in S_i are serviced until the head buffer, $(j, 1)$, $j \in S_i$, of every path in S_i is clear. At that point, the head buffers are no longer serviced and the production run continues until all buffers, except possibly the head buffers, in the paths of S_i are empty. In the following analysis, we will assume that all buffers except the head buffers are initially empty (if this is not true initially, it will be after each path has been serviced once). For convenience, we will refer to the level of the head buffer of path i, $(i, 1)$ at time t as $x_i(t)$. The goal of the following analysis is to identify GSC policies which are stable (i.e., buffer levels remain bounded for all time) and to obtain bounds on the buffers of FMS controlled by such policies.

5.4.2 Global Synchronous Clear-A-Fraction Policy

The first GSC policy that we consider is the GSCAF (global synchronous clear-a-fraction) policy. This policy is a generalization of the local CAF policy for single machines, introduced in [86]. We define the GSCAF policy by specifying how it chooses the next set of paths to service upon the ending of a production run. Define a function $V(X(t))$ as

$$V(X(t)) \triangleq \sum_{i=1}^{N} P_i x_i(t) . \tag{5.85}$$

Notice that $V(X(t))$ is the sum of all parts in the head buffers of all paths in the network (at the end of a production run, there can be no parts in buffers that are not head buffers) weighted by the number of paths that the parts must traverse, after and including the path that they are currently in, before exiting the FMS. If a scheduling decision is to be made at time t, the GSCAF policy chooses to service a set of paths S_i, such that

$$\sum_{j \in S_i} x_j(t) \geq \epsilon V(X(t)) ,$$

for some sufficiently small, fixed $\epsilon > 0$. We must choose ϵ small enough so that regardless of the distribution of parts at the end of any production run, there is some $i \in \{1, 2, \ldots, M\}$ that satisfies the above selection criterion. It is easy to identify $\epsilon^* > 0$, such that for any $\epsilon \leq \epsilon^*$ there is some $i \in \{1, 2, \ldots, M\}$ such that the selection criterion is satisfied. For instance,

$$\epsilon^* = \frac{1}{N\bar{P}},$$

where $\bar{P} = \max_i P_i$, will work.

In summary, during a production run, the GSCAF policy completely clears all parts from each path processed (except for parts which may accumulate at the head buffers after processing has stopped at the head buffers). Notice that because all buffers in a path are serviced at the minimum processing rate for the path, parts never accumulate at any buffers in the path other than the head buffer. In other words, along each individual path, parts arrive at downstream buffers at a rate that is no greater than the rate at which they can be processed. This is sometimes referred to as "pipelining" parts. Next, we shall develop conditions under which the GSCAF policy is stable.

To begin with, we define several parameters that will be used in the stability analysis. First, define a set of times $T = \{t_0, t_1, t_2, \ldots\}$, where t_0 is the FMS start-up time and beginning of the first production run, t_1 is the end of the first production run and beginning of the second production run, and, in general, t_i is the end of production run i and the beginning of production run $i+1$. Let $\max\{s_j : j \in S_i\}$ be denoted by $\bar{s}^i$, $\max\{D_j : j \in S_i\}$ by $\bar{D}^i$, $\max\{\tau_j : j \in S_i\}$ by $\bar{\tau}^i$, $\min\{\tau_j : j \in S_i\}$ by $\underline{\tau}^i$, $\max\{a_j : j \in S_i\}$ by $\bar{a}^i$, $\max\{\bar{s}^i : 1 \leq i \leq M\}$ by $\bar{s}$, and $\max\{\bar{a}^i : 1 \leq i \leq M\}$ by $\bar{a}$. Let us choose an arbitrary time $t_p \in T$. Let us assume that the GSCAF policy has chosen to service buffers in the paths of S_i during the production run beginning at time t_p. For every $i \in \{1, 2, \ldots, M\}$, define a constant η_i, $0 < \eta_i \leq 1$, such that, at any time t_p,

$$\max\{x_j(t_p) : j \in S_i\} \quad \leq \quad \eta_i \sum_{j \in S_i} x_j(t_p) \, . \tag{5.86}$$

Notice that η_i follows from the topology of the FMS. In many cases, we may have to take $\eta_i = 1$ because $|S_i| = 1$ or because the topology of the FMS does not allow for $\eta_i < 1$ in general (note that if an $\eta_i < 1$ cannot be found then $\eta_i = 1$ will suffice for the stability analysis below, no matter what FMS topology is being considered). As an example of a situation in which we can take $\eta_i < 1$, consider paths $S_i = \{m, n\} \subset Q$, where $m \notin S_j$ and $n \notin S_j$ for all $j \in \{1, 2, \ldots, M\}$, $j \neq i$. Because paths m and n were both last cleared at the same time, $x_m(t_p) = a_m t'$ and $x_n(t_p) = a_n t'$, for some $t' > 0$. From the above definition of η_i, we see that

$$\eta_i = \max\left\{\frac{x_m(t_p)}{x_m(t_p) + x_n(t_p)}, \frac{x_n(t_p)}{x_m(t_p) + x_n(t_p)}\right\}.$$

Hence,

$$\eta_i \; = \; \max\left\{\frac{a_m}{a_m + a_n}, \frac{a_n}{a_m + a_n}\right\} \, .$$

Generally, it is desirable to find an η_i that is as small as possible as the conditions for stability then become less restrictive and the bounds obtained on buffer levels become tighter, as can be seen by the following stability result.

Theorem 5.7 (Global Synchronous Clear-A-Fraction Policy): *Let $F(t)$ be a function that returns the integer number of production runs that have been completed during the time interval $[t_0, t]$. If, for all $i \in \{1, 2, \ldots, M\}$,*

$$\eta_i \frac{\sum_{j \notin S_i} P_j a_j + \sum_{j \in S_i} (P_j - 1) a_j}{\underline{\tau}^i - \bar{a}^i} \quad < \quad 1 ,$$

then the GSCAF policy is stable. Furthermore,

$$\sum_{i=1}^{N} P_i x_i(t) \quad \leq \quad \gamma^{F(t)} \left(V(X(t_0)) - \frac{\zeta}{1-\gamma} \right) + \frac{\zeta}{1-\gamma} + \sum_{i=1}^{N} P_i \bar{s} \bar{a} ,$$

where

$$\zeta_i \quad = \quad \frac{\bar{s}^i + \bar{D}^i}{1 - \frac{\bar{a}^i}{\underline{\tau}^i}} \left[\sum_{j \notin S_i, j \in Q} P_j a_j + \sum_{j \in S_i} (P_j - 1) a_j \right] + \sum_{j \in S_i} P_j \bar{D}^i a_j ,$$

$$\gamma_i \quad = \quad 1 - \epsilon \left[1 - \eta_i \frac{\sum_{j \notin S_i} P_j a_j + \sum_{j \in S_i} (P_j - 1) a_j}{\underline{\tau}^i - \bar{a}^i} \right]$$

and $\gamma = \max\{\gamma_i\}$, $\zeta = \max\{\zeta_i\}$.

Proof: We first bound the length of the production run beginning at t_p in terms of $\sum_{j \in S_i} x_j(t_p)$ by a value which we shall call Δ_p. There are four factors which are summed to form Δ_p: (a) a bound on the maximum amount of time to clear parts in any head buffer of a path in S_i,

$$\frac{\eta_i \sum_{j \in S_i} x_j(t_p)}{\underline{\tau}^i},$$

(b) a bound on the maximum amount of time to clear parts which arrive at any head buffer of a path in S_i during the production run, $\bar{a}^i \Delta_p / \underline{\tau}^i$, (c) the maximum set-up time for any buffer in the paths in S_i, $\bar{s}^i$, and (d) the maximum total transport time of any of the paths in S_i, $\bar{D}^i$. Hence,

$$\Delta_p \quad \leq \quad \frac{\bar{a}^i \Delta_p + \eta_i \sum_{j \in S_i} x_j(t_p)}{\underline{\tau}^i} + \bar{s}^i + \bar{D}^i \tag{5.87}$$

$$= \quad \frac{\eta_i \frac{\sum_{j \in S_i} x_j(t_p)}{\underline{\tau}^i} + \bar{s}^i + \bar{D}^i}{1 - \frac{\bar{a}^i}{\underline{\tau}^i}} . \tag{5.88}$$

Next we bound $V(X(t_{p+1}))$ in terms of $V(X(t_p))$. In order to accomplish this, we consider the contribution to V by several subsets of S. First, notice that

$$\sum_{j \in S_i} P_j x_j(t_{p+1}) \quad \leq \quad \sum_{j \in S_i} P_j x_j(t_p) - \sum_{j \in S_i} P_j x_j(t_p) + \sum_{j \in S_i} P_j \bar{D}^i a_j . \tag{5.89}$$

This is because all paths in S_i are cleared during the production run and because some parts may accumulate at the head buffers during any part transport time before the end of the production run. Next, notice that

$$\sum_{j \notin S_i, j \in Q} P_j x_j(t_{p+1}) \leq \sum_{j \notin S_i, j \in Q} P_j x_j(t_p) + \sum_{j \notin S_i, j \in Q} P_j \Delta_p a_j . \quad (5.90)$$

This is because all paths not in S_i that are fed by network inputs have their buffer levels grow at the rate of part arrival during the production run. Let $H^*_{S_i}$ be the set of all paths fed by paths in S_i so that $H^*_{S_i} = \{j = H_k : k \in S_i, P_k > 1\}$. Then we see that

$$\sum_{j \in H^*_{S_i}} P_j x_j(t_{p+1}) \leq \sum_{j \in H^*_{S_i}} P_j x_j(t_p) + \sum_{j \in S_i} (P_{j-1}) \left[x_j(t_p) + \Delta_p a_j \right] . \quad (5.91)$$

This is because a path which is fed by some path $j \in S_i$ has its head buffer increase during the production run by no more than $x_j(t_p) + \Delta_p a_j$, a bound on the total number of parts processed through path j during the production run. Notice that the P_{j-1} term in the final sum is zero if path j feeds a network output. Finally, notice that

$$\sum_{\substack{j \notin S_i, j \in R \\ j \notin H^*_{S_i}}} P_j x_j(t_{p+1}) = \sum_{\substack{j \notin S_i, j \in R \\ j \notin H^*_{S_i}}} P_j x_j(t_p) . \quad (5.92)$$

This is because paths not in S_i that are fed neither by network inputs nor paths in S_i have their head buffers remain unchanged during the production run. If we sum the four preceding (in)equalities, we see that

$$\begin{aligned} V(X(t_{p+1})) \leq\ & V(X(t_p)) - \sum_{j \in S_i} P_j x_j(t_p) \\ & + \sum_{j \in S_i} P_j \bar{D}^i a_j + \sum_{j \notin S_i} P_j \Delta_p a_j \\ & + \sum_{j \in S_i} (P_j - 1) \left[x_j(t_p) + \Delta_p a_j \right] . \end{aligned} \quad (5.93)$$

Manipulating the above expression and using the bound on Δ_p, we see that an upper bound on $V(X(t_{p+1}))$ is

$$V(X(t_p)) - \sum_{j \in S_i} x_j(t_p)$$

$$+ \left(\frac{\eta_i \sum_{j \in S_i} x_j(t_p)}{\underline{\tau}^i - \bar{a}^i} + \frac{\bar{s}^i + \bar{D}^i}{1 - \frac{\bar{a}^i}{\underline{\tau}^i}} \right) \left[\sum_{j \notin S_i} P_j a_j + \sum_{j \in S_i} (P_j - 1) a_j \right]$$

$$
\begin{aligned}
&+\sum_{j\in S_i} P_j \bar{D}^i a_j && (5.94)\\
&= V(X(t_p)) + \zeta_i - \\
&\left(\sum_{j\in S_i} x_j(t_p)\right)\left[1-\eta_i\frac{\sum_{j\notin S_i} P_j a_j + \sum_{j\in S_i}(P_j-1)a_j}{\underline{\tau}^i - \bar{a}^i}\right] && (5.95)\\
&\le V(X(t_p)) + \zeta_i - \\
&\epsilon V(X(t_p))\left[1-\eta_i\frac{\sum_{j\notin S_i} P_j a_j + \sum_{j\in S_i}(P_j-1)a_j}{\underline{\tau}^i - \bar{a}^i}\right] && (5.96)\\
&= \gamma_i V(X(t_p)) + \zeta_i\ , && (5.97)
\end{aligned}
$$

where

$$
\zeta_i = \frac{\bar{s}^i + \bar{D}^i}{1-\frac{\bar{a}^i}{\underline{\tau}^i}}\left[\sum_{j\notin S_i} P_j a_j + \sum_{j\in S_i}(P_j-1)a_j\right] + \sum_{j\in S_i} P_j \bar{D}^i a_j \qquad (5.98)
$$

and

$$
\gamma_i = 1-\epsilon\left[1-\eta_i\frac{\sum_{j\notin S_i} P_j a_j + \sum_{j\in S_i}(P_j-1)a_j}{\underline{\tau}^i - \bar{a}^i}\right]. \qquad (5.99)
$$

Due to the sufficient condition of Theorem 5.7, $0 < \gamma_i < 1$. Notice that the sufficient condition of Theorem 5.7 is satisfied if $\underline{\tau}^i$ is sufficiently large with respect to $\bar{a}^i$ for all $i \in \{1, 2, \ldots, M\}$. If we let $\gamma = \max\{\gamma_i : i = 1, 2, \ldots, M\}$ and $\zeta = \max\{\zeta_i : i = 1, 2, \ldots, M\}$, we see that for any time $t_p \in T$,

$$
V(X(t_{p+1})) \le \gamma V(X(t_p)) + \zeta\ . \qquad (5.100)
$$

We can solve this first order difference inequality to yield

$$
V(X(t_k)) \le \gamma^k\left(V(X(t_0)) - \frac{\zeta}{1-\gamma}\right) + \frac{\zeta}{1-\gamma}\ .
$$

Choose any $t_p \in T$ such that the buffers of path S_i are serviced during the production run beginning at time t_p. It is clear that in the closed time interval $[t_p, t_{p+1}]$, the maximum of $V(X(t))$ occurs either at t_p, t_{p+1}, or at a point t', which occurs sometime during the span of time during which servicing begins on the head buffers of paths in S_i. Specifically, $t_p + \underline{s}^i \le t' \le t_p + \bar{s}^i$. Consequently, if we let $\bar{s} = \max\{\bar{s}^i : i = 1, 2, \ldots M\}$, we see that for all $t \ge t_0$,

$$
V(X(t)) \le \gamma^{F(t)}\left(V(X(t_0)) - \frac{\zeta}{1-\gamma}\right) + \frac{\zeta}{1-\gamma} + \sum_{i\in Q} P_i \bar{s} a_i\ . \qquad (5.101)
$$

Further, we see from the definition of $V(X(t))$ that

$$\sum_{i=1}^{N} P_i x_i(t) \leq \gamma^{F(t)} \left(V(X(t_0)) - \frac{\zeta}{1-\gamma} \right) + \frac{\zeta}{1-\gamma} + \sum_{i=1}^{N} P_i \bar{s} \bar{a}, \tag{5.102}$$

for all $t \geq t_0$. ■

Notice that while the condition for the stability of the GSCAF policy is not as intuitive as the capacity condition of the CAF policy, it is, in some sense, a generalization of the capacity condition because, for each $i \in \{1, 2, \ldots, M\}$, it requires that the rate of growth of some fraction (η_i) of the growth of V during a production run be less than the rate at which V is decreased during the production run.

Next, notice that the GSC policies do not truly require global information (i.e., all buffer levels and the status of processing at each machine). Because of the manner in which a GSC policy mandates part pipelining by synchronizing production in consecutive machines, the policy only needs to know (a) when each head buffer has been cleared during a production run (so that the policy can stop production at the head buffers when they all have been cleared) (b) when all paths currently being serviced are empty downstream from the head buffers (so that the policy can begin a new production run), and (c) the buffer levels of all head buffers at the end of each production run (so that the policy can select a new set of paths to service).

Similar to the way in which the clear-largest-buffer (CLB) policy is a special case of policy CAF [86], we can define the global synchronous clear-largest-buffer (GSCLB) policy as a special case of GSCAF. The GSCLB policy chooses to process the paths in the set S_i, $i \in \{1, 2, \ldots, M\}$, with the largest sum of parts. More precisely, GSCLB chooses at time t_p to process paths of some set S_i, $i \in \{1, 2, \ldots, M\}$, such that

$$\sum_{j \in S_i} x_j(t_p) \geq \sum_{j \in S_q} x_j(t_p), \tag{5.103}$$

for all $q \in \{1, 2, \ldots, M\}$, $q \neq i$. GSCLB is considered to be a special case of GSCAF because at each time GSCLB picks an S_i to process, this S_i could have also been chosen by GSCAF. Due to this fact, Theorem 5.7 also applies to the GSCLB policy.

The primary drawback of the above analysis is the conservative nature of the sufficient condition for stability. The fact is that many FMS which do not satisfy the condition are indeed stable. In the next section, we consider a priority GSC policy that for a large number of FMS topologies will alleviate the conservative nature of the stability condition and produce much sharper buffer bounds than the GSCAF analysis.

5.4.3 Global Synchronous Periodic Clearing Policy

The scheduler that we now describe is inspired by the observation that when a GSC policy is allowed to control an FMS, it often falls into a periodic pattern of servicing choices. This periodicity seems to appear whether or not the resulting FMS is stable. Hence, what we call the "Global Synchronous Periodic Clearing" (GSPC) Policy will enforce a given, periodic service schedule. There are several ways to view such a policy. In some cases, intuition may be used to specify a periodic policy which will perform well. In other cases, service periods which perform well may be identified by simulation or actual implementation. Finally, we may identify a stable service period and use it as a "stabilizing safety net" for some other GSC policy that we cannot rigorously guarantee is stable.

For our stability analysis, we begin by specifying a special model of the FMS being considered. Let $y(p)$ be the vector of the head buffer levels at time t_p so that

$$\begin{aligned} y(p) &= [y_1(p), y_2(p), \ldots, y_N(p)]^\top \\ &= [x_1(t_p), x_2(t_p), \ldots, x_N(t_p)]^\top \end{aligned} \tag{5.104}$$

The key to our analysis in this section is the fact that we can write

$$y(p+1) = A_{i,j}y(p) + b_{i,j} \tag{5.105}$$

for some $A_{i,j} \in \Re^{N\times N}_{[0,\infty)}$ and $b_{i,j} \in \Re^N$ ($\Re^{N\times N}_{[0,\infty)}$ is the set of all real $N \times N$ matrices with nonnegative entries). The subscript "i, j" in the above expression denotes that paths in the set S_i are serviced during the production run beginning at time t_p and that path $j \in S_i$ is a *critical path*. In calling path j a "critical path," we mean that it is last of the paths in S_i to have its head buffer cleared during the production run (i.e., when the head buffer of path j is cleared, processing ceases at all of the head buffers of paths in S_i). Let n_i be the number of possible critical paths in S_i. If j is the critical path (ties are resolved arbitrarily) and paths in S_i are serviced during the production run beginning at time t_p, we can write Δ_p, the time it takes to complete the production run begun at time t_p, as

$$\begin{aligned} \Delta_p &= \frac{y_j(p)}{\tau_j} + \frac{a_j(\Delta_p - \bar{D}^i)}{\tau_j} + s_j + \bar{D}^i \\ &= \frac{\frac{y_j(p) - a_j\bar{D}^i}{\tau_j} + s_j + \bar{D}^i}{1 - \frac{a_j}{\tau_j}} \\ &= \frac{y_j(p)}{\tau_j - a_j} + \frac{s_j + \bar{D}^i - \frac{a_j\bar{D}^i}{\tau_j}}{1 - \frac{a_j}{\tau_j}} . \end{aligned} \tag{5.106}$$

Notice that this expression for Δ_p is similar, but not exactly the same as, the expression for Δ_p in the GSCAF analysis. The difference arises because this is an exact expression for Δ_p, where the expression in the GSCAF analysis is an upper bound.

Let $[A_{i,j}]_{m,n}$ denote the element in row m and column n of $A_{i,j}$, and let $[b_{i,j}]_m$ denote element m of $b_{i,j}$. We now describe how to build $A_{i,j}$ and $b_{i,j}$ in Equation (5.105) for an FMS of arbitrary topology. First, consider $m \in S_i$. Because all paths in S_i are cleared, except for parts which arrive in the time segment of length $\bar{D}^i$ between the end of head buffer servicing and the end of the production run, we see that

$$[A_{i,j}]_{m,n} = 0, \; n \in \{1, 2, \ldots, N\} \tag{5.107}$$

and

$$[b_{i,j}]_m = \bar{D}^i a_m . \tag{5.108}$$

Secondly, consider $m \notin S_i$, $m \in Q$ (i.e., path m is not being serviced and is fed by a network input). Clearly, $y_m(p+1) = y_m(p) + a_m \Delta_p$, so that if j is the critical path and we use Equation (5.106),

$$[A_{i,j}]_{m,n} = \begin{cases} 1 & n = m \\ \frac{a_m}{t_j - a_j} & n = j \\ 0 & \text{otherwise} \end{cases} . \tag{5.109}$$

and

$$[b_{i,j}]_m = a_m \frac{s_j + \bar{D}^i - \frac{a_j \bar{D}^i}{\tau_j}}{1 - \frac{a_j}{\tau_j}} . \tag{5.110}$$

Thirdly, consider $m \in H^*_{S_i}$ with $H_q = m$, $q \in S_i$ (i.e., path m is not being serviced, but is fed by path q that is being serviced). Because m is fed by $q \in S_i$, we see that $y_m(p+1) = y_m(p) + y_q(p) + a_q(\Delta_p - \bar{D}^i)$. Hence, if j is the critical path and we use Equation (5.106),

$$[A_{i,j}]_{m,n} = \begin{cases} 1 & n = m \\ 1 & n = q \text{ and } q \neq j \\ \frac{a_q}{t_j - a_j} & n = j \text{ and } q \neq j \\ 1 + \frac{a_j}{t_j - a_j} & n = q \text{ and } q = j \\ 0 & \text{otherwise} \end{cases} \tag{5.111}$$

(note that the second and third lines correspond to the case in which the path that feeds path m is not the critical path and that the fourth line corresponds to the case in which the path which feeds path m is the critical path), and

$$[b_{i,j}]_m = a_q \left(\frac{s_j + \bar{D}^i - \frac{a_j \bar{D}^i}{\tau_j}}{1 - \frac{a_j}{\tau_j}} - \bar{D}^i \right) . \tag{5.112}$$

Finally, consider $m \notin S_i$, $m \in R$, and $m \notin H^*_{S_i}$ (i.e., path m is not being serviced and is fed neither by a network input nor by a path which is being serviced). Because path m is fed neither by a network input nor by some path in S_i, it is clear that $y_m(p+1) = y_m(p)$, so that

$$[A_{i,j}]_{m,n} = \begin{cases} 1 & n = m \\ 0 & \text{otherwise} \end{cases}, \tag{5.113}$$

and

$$[b_{i,j}]_m = 0 . \tag{5.114}$$

This completes the definition of the model in Equation (5.105) for the FMS.

Before we begin our analysis, we list three results from linear algebra that we shall find useful. In the following, we shall let $\|\cdot\|$ denote the vector two-norm or the matrix norm induced by the vector two-norm:

$$\|x\| = \left(\sum_{i=1}^{N} x_i^2 \right)^{1/2}, \quad x \in \Re^N \tag{5.115}$$

$$\|A\| = \max_{\|x\|=1} \|Ax\| , \quad A \in \Re^{N \times N} , \tag{5.116}$$

and we shall let $\lambda(A)$ denote the set of eigenvalues of the matrix $A \in \Re^{N \times N}$.

Lemma 1: Let $A \in \Re^{N \times N}$. Then $\lim_{k \to \infty} A^k = 0$ if and only if

$$\max(|\lambda(A)|) < 1.$$

Lemma 2: Let $A \in \Re^{N \times N}$ and $\|A\|_M < 1$, where $\|\cdot\|_M$ is any matrix norm. Then

$$\sum_{k=0}^{\infty} A^k = (I - A)^{-1} . \tag{5.117}$$

Lemma 3: Let $A \in \Re^{N \times N}$. If $\max(|\lambda(A)|) < 1$, then there is some matrix norm $\|\cdot\|_M$ such that $\|A\|_M < 1$.

Next, we formally specify the GSCP policy. To do this, we specify an ordered list, ρ, of length N_ρ, whose members are elements of $\{1, 2, \ldots, M\}$. The elements of ρ are referenced by $\rho(1), \rho(2), \ldots, \rho(N_\rho)$. The list ρ specifies one period of a periodic servicing sequence. For example, once the GSPC policy is activated, the periodic servicing sequence is $S_{\rho(1)}, S_{\rho(2)}, \ldots, S_{\rho(N_\rho)}$,

$S_{\rho(1)}, S_{\rho(2)}, \ldots, S_{\rho(N_\rho)}, \ldots$. Of course, our intuition tells us that a necessary condition for the periodic servicing sequence to produce stable operation is that the sequence contain at least one occurrence of each S_i, $i \in \{1, 2, \ldots, M\}$.

Let the notation

$$\prod_{i=n,-1}^{m} A_{\phi(i)} , \quad n \geq m , \tag{5.118}$$

where $A_{\phi(i)}$ denotes a particular square matrix for all $i \in \{m, m+1, \ldots, n\}$, denote the product

$$A_{\rho(n)} A_{\phi(n-1)} \cdots A_{\phi(m+1)} A_{\phi(m)} . \tag{5.119}$$

Let t_0 be the time at which the GSPC policy is initiated. Using the model developed above, we can write

$$y(1) = A_{\rho(1),j_0} y(0) + b_{\rho(1),j_0} \tag{5.120}$$

$$y(2) = A_{\rho(2),j_1} y(1) + b_{\rho(2),j_1} \tag{5.121}$$

$$= A_{\rho(2),j_1} A_{\rho(1),j_0} y(0) + A_{\rho(2),j_1} b_{\rho(1),j_0} + b_{\rho(2),j_1} \tag{5.122}$$

$$\vdots$$

$$y(N_\rho) = \left(\prod_{i=N_\rho,-1}^{1} A_{\rho(i),j_{i-1}} \right) y(0) + b_{\rho(N_\rho),j_{N_\rho - 1}} + \sum_{i=1}^{N_\rho - 1} \left(\prod_{k=N_\rho,-1}^{i+1} A_{\rho(k),j_{\rho(k)-1}} \right) b_{\rho(i),j_{\rho(i)-1}} , \tag{5.123}$$

where j_i denotes the critical path for the production run beginning at time t_i.

There are

$$n_\rho \triangleq n_{\rho(1)} n_{\rho(2)} \cdots n_{\rho(N_\rho)} \tag{5.124}$$

ways in which the functions of matrices

$$\prod_{i=N_\rho,-1}^{1} A_{\rho(i),j_{i-1}} \tag{5.125}$$

and

$$b_{\rho(N_\rho),j_{N_\rho - 1}} + \sum_{i=1}^{N_\rho - 1} \left(\prod_{k=N_\rho,-1}^{i+1} A_{\rho(k),j_{\rho(k)-1}} \right) b_{\rho(i),j_{\rho(i)-1}} \tag{5.126}$$

might be formed. Notice that n_i is the number of different critical paths that can feasibility result when paths in S_i are serviced. We form the sets

$$A^* \triangleq \{A_1^*, A_2^*, \ldots, A_{n_\rho}^*\} \quad \text{and} \quad b^* \triangleq \{b_1^*, b_2^*, \ldots b_{n_\rho}^*\} \tag{5.127}$$

to contain all possible formations of the matrix functions. This completes the definition of the GSPC policy and the associated FMS model.

In what follows, we derive stability conditions and buffer bounds for FMS that satisfy different topological conditions. We first consider an important class of systems with the topological property that $n_i = 1$ for all i, $1 \leq i \leq M$. Notice that this condition is satisfied by systems for which $|S_i| = 1$ for all i, $1 \leq i \leq M$, and that it is also effectively satisfied by systems in which any S_i with $|S_i| > 1$ contains only paths which are fed by network inputs. The last part of the previous statement is true because in set S_i with $|S_i| > 1$ which contain only paths which are fed by network inputs, after the initial production run, the critical path for all future production runs is determined by the fixed input and processing rates of the paths of S_i. Two examples of FMS with $n_i = 1$ are the highly re-entrant line and the feedforward line of Section 5.5 (see Figures 5.4 and 5.5). In all systems with $n_i = 1$ for all i, $1 \leq i \leq M$, the sets A^* and b^* each contain just one element, A_1^* and b_1^*, respectively. Let $A \triangleq A_1^*$ and $b \triangleq b_1^*$. In terms of A and b, then, we have the iterative relation

$$y((k+1)N_\rho) = Ay(kN_\rho) + b \ . \tag{5.128}$$

Theorem 5.8 (Global Synchronous Periodic Clearing Policy: Case 1): *If $n_i = 1$ for all i, $1 \leq i \leq M$, and if $\max(|\lambda(A)|) < 1$, then the GSPC-controlled FMS is stable. Furthermore,*

$$y(kN_\rho) = A^k y(0) + \left(\sum_{i=0}^{k-1} A^i\right) b \tag{5.129}$$

and

$$\lim_{k\to\infty} y(kN_\rho) = (I - A)^{-1} b \tag{5.130}$$

Proof: That $y(kN_\rho)$ can be written as

$$y(kN_\rho) = A^k y(0) + \left(\sum_{i=0}^{k-1} A^i\right) b \tag{5.131}$$

is easily shown via induction on the iterative relation $y((k+1)N_\rho) = Ay(kN_\rho) + b$. Given this, it is easy to see that

$$\lim_{k\to\infty} y(kN_\rho) = (I - A)^{-1} b \ , \tag{5.132}$$

because (a) $\lim_{k\to\infty} A^k = 0$ by Lemma 1 and (b) $\sum_{i=0}^{\infty} A^i = (I - A)^{-1}$ by Lemmas 2 and 3. Hence, the system is stable because we have

$$\|(I - A)^{-1}\| \|b\| < \infty.$$

This completes the proof. ■

Notice that the "bounds" provided by this result are actually exact characterizations of FMS behavior. In this case, we do not need to resort to norm-based buffer bounds. Notice also that the result is only valid once in every N_ρ times. In order to find the behavior at all times k, we must form N_ρ different A and b matrices and compute the behavior for each set. Each of the A and b matrix sets should correspond to the length-N_ρ service period starting at a different point. For example, if the original service period is S_1, S_2, S_3, we need to calculate three different sets of A and b matrices, with one corresponding to each of the following periods: $S_1, S_2, S_3,$ S_2, S_3, S_1, and S_3, S_1, S_2.

Next, notice that it is always possible to define the S_i so that $n_i = 1$ (e.g. choose the S_i, so that $|S_i| = 1$ for all i). However, from a performance (and stability) perspective, this is clearly not always the best choice. For example, an FMS that, due to its topology, requires the defining of many short paths (such as the cellular structure I FMS of Section 5) may perform badly or even become unstable if we take $|S_i| = 1$ for all i. These types of FMS generally require more than a single path to be processed at once.

We are now left to consider systems with $n_i > 1$ for some i, $1 \leq i \leq M$ (see the cellular structure examples of Section 5.5 in Figures 5.6 and 5.7). As you will see below, it is unfortunately that case that the bounds we obtain are not nearly so sharp as the bounds in the $n_i = 1$ case.

Before presenting the stability result, we define some convenient notation. For a fixed product of matrices from the set A^*,

$$\prod_{i=r,-1}^{1} A^*_{\phi(i)}\,, \tag{5.133}$$

where $\phi(i)$ is an index into the set A^*, we define the *corresponding sum of products* to be the vector b that is appropriate in the iterative relation

$$y((k+1)rN_\rho) = \left(\prod_{i=r,-1}^{1} A^*_{\phi(i)}\right) y(krN_\rho) + b\,. \tag{5.134}$$

In other words,

$$b = b^*_{\phi(r)} + \sum_{i=1}^{r-1}\left(\prod_{j=r,-1}^{i+1} A^*_{\phi(j)}\right) b^*_{\phi(i)}\,. \tag{5.135}$$

Theorem 5.9 (Global Synchronous Periodic Clearing Policy: Case 2): *Suppose that there exists integer $r > 0$ such that any product of r matrices from the set A^* has matrix two-norm less than 1. Let A denote the r-length product with the largest norm, and let b denote the* *corresponding sum of products. The GSPC-controlled FMS is stable and*

$$||y(krN_\rho)|| \leq ||A||^k ||y(0)|| + \sum_{i=0}^{k-1} ||A||^i ||b|| \qquad (5.136)$$

so that

$$\lim_{k \to \infty} ||y(krN_\rho)|| \leq \frac{||b||}{1 - ||A||}. \qquad (5.137)$$

Proof: From the definitions of A and b, it is clear that we can write

$$||y((k+1)rN_\rho)|| \leq ||A|| ||y(krN_\rho)|| + ||b||. \qquad (5.138)$$

By applying induction to the above iterative relation, it easily follows that

$$||y(krN_\rho)|| \leq ||A||^k ||y(0)|| + \sum_{i=0}^{k-1} ||A||^i ||b||. \qquad (5.139)$$

Because we have required that $||A|| < 1$, it is also apparent that

$$\lim_{k \to \infty} ||y(krN_\rho)|| \leq \frac{||b||}{1 - ||A||} < \infty. \qquad (5.140)$$

This completes the proof. ■

The conditions of Theorem 5.9 are not completely satisfying for two reasons. First of all, we must check all length-r multiplicative combinations of the A_i^* matrices, regardless of whether all of the combinations are physically realizable. Secondly, we would like to avoid a norm-based result. Next, notice that even though the GSPC policy prescribes a periodic service schedule, it is still a feedback policy in that it must know when production runs have been completed. Also note that for either $n_i = 1$ or $n_i > 1$, it may be the case that the production engineer may not be aware of a good choice of service period for the GSPC policy. In this case, one approach is to identify a stable service period and use it as a supervisor for another, more intuitive (e.g., GSCLB) policy. As a supervisor, the purpose of the GSPC policy is simply to guarantee stability of the system. In practice, the supervising GSPC policy is invoked when some measure of system performance (e.g., the sum of the buffer levels) exceeds some preset threshold (i.e., once the sum of FMS buffer levels grows beyond some pre-set limit, a pre-defined, stable periodic service sequence is implemented).

5.5 SCHEDULING EXAMPLES: CENTRALIZED VS. DISTRIBUTED

In the final analysis, the main reason that we choose to use a particular scheduling policy to control a particular FMS is because it performs well. The point that we attempt to make in this section, via simulation, is that for some FMS GSC policies can perform much better than other distributed scheduling policies, and that for other FMS, some distributed policies exhibit better performance (distributed policies are isolated, local policies such as the CAF policy of [86] that are supervised by the USSM of [42]).

In attempting to formulate a general rule-of-thumb for determining the suitability of GSC control for a given FMS, we make the following observation: The less variance there is among processing rates along individual paths, the better GSC policies will perform (with respect to distributed policies). The reason for this is that because GSC policies mandate that all buffers on a given path be processed at a single rate (the minimum processing rate of all buffers on the path), any buffers on the path that are able to be processed at a faster rate than the minimum processing rate are constrained to be processed at a lower rate than they would be processed at in a distributed control scheme. In general, then, for systems with very high processing rate "skew" along individual paths, we may be wiser to choose a distributed policy. However, it may be possible to choose paths intelligently so as to minimize the adverse affects of processing rate skew.

In the following examples, we compare the performance of GSC policies to that of distributed policies. We will look at several different topologies, and we will vary the FMS parameters from those favoring GSC policies to those favoring distributed policies. In all of the simulations, we will take all network part transportation delay times to be zero. As a performance metric, we choose average backlog of all part types, Ψ. In order to precisely define Ψ, we need to define a new buffer indexing scheme. For a given FMS, let there be n different part types. Accordingly, define the n sets $L_1, L_2, \ldots, L_n$, where L_i, $i \in \{1, 2, \ldots, n\}$, contains the indices of all buffers through which parts of type i pass. In order to avoid confusion with our previous buffer indexing scheme, define $\xi_i(t)$ to be the level of the buffer referred to by index i (and therefore a buffer which contains parts of type j, $i \in L_j$) at time $t \geq t_0$. We define the average backlog of parts of type i, ψ_i, as

$$\psi_i = \sum_{j \in L_i} \lim_{t \to \infty} \left(\frac{1}{t - t_0} \int_{t_0}^{t} \xi_j(\alpha) d\alpha \right), \tag{5.141}$$

and we define Ψ as the average of the ψ_i,

$$\Psi = \frac{1}{n} \sum_{i=1}^{n} \psi_i . \tag{5.142}$$

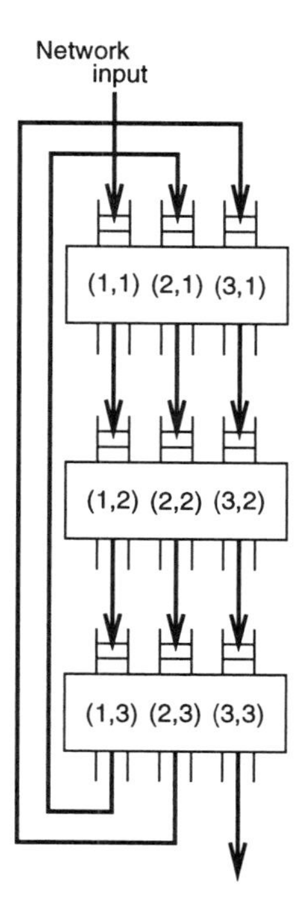

Fig. 5.4 A highly re-entrant line.

In order to approximate the infinite-time averages required in the above expression, we will run the simulations for 1000 production runs and take the averages over the last 500 (longer runs produce nearly identical results).

5.5.1 Highly Re-entrant Line

The re-entrant line form of an FMS is common in industry (particularly in the semi-conductor fabricating industry) and is considered important enough to have warranted studies specifically aimed at analyzing its specialized topology (e.g., see [42, 41]). As an example of a highly re-entrant line, we will simulate the FMS of Figure 5.4. In the FMS of Figure 5.4, the buffers have been labeled according to the convention that we outlined earlier; path 1 consists of buffers $(1,1)$, $(1,2)$, and $(1,3)$, path 2 consists of buffers $(2,1)$, $(2,2)$, and $(2,3)$, and path 3 consists of buffers $(3,1)$, $(3,2)$, and $(3,3)$. Accordingly, we choose $S_1 = \{1\}$, $S_2 = \{2\}$, and $S_3 = \{3\}$.

We will perform three simulations of the re-entrant line of Figure 5.4 with three different sets of processing rate parameters. For each simulation, we will fix the sole network arrival rate, a_1, to be $4/3$, and we will assume that all buffers have a set-up time of 50.0. The processing rate parameters that

Table 5.4 Buffer Processing Rates

	Simulation		
Buffer	1	2	3
(1,1)	5.0	3.3	20
(1,2)	5.0	4.0	5
(1,3)	5.0	10.0	20
(2,1)	5.0	10.0	5
(2,2)	5.0	5.0	20
(2,3)	5.0	5.0	20
(3,1)	5.0	20.0	20
(3,2)	5.0	10.0	20
(3,3)	5.0	20.0	5

we will use for each simulation are listed in Table 5.4. Notice that the first simulation has processing rates that are uniform along the paths, the second simulation has processing rates that are quite skewed along the paths, and the third simulation has processing rates that are extremely skewed along the paths.

As a representative distributed policy, we will use distributed CLB, as it has always performed better than CAF and the "CPK policy" (clearing policy of Perkins and Kumar) of [86] for this type of topology (for a simulation-based comparison of CLB, CAF, and CPK, see [1]). For an in-depth look at the CAF and CLB policies, see [86]. In order to guarantee stability of the distributed CLB policy, we employ the USSM stabilizing mechanism of [43] with the parameters set large enough that the USSM never intervenes. For our GSC policy, we will choose GSCLB. In order to guarantee stability of our GSC policy, we will supervise the GSCLB policy with the GSPC policy which enforces the service period S_1, S_2, S_3 if the buffer levels ever grow too large (which never happens in any of the simulations). This GSPC policy can easily be shown to be stable by Theorem 5.8 for each of the three simulation parameter sets.

Table 5.5 summarizes the results of the three simulations. Notice that the GSCLB policy performs the same in the first and third simulations. This is because in both cases, the minimum processing rate on all three paths is 5. In the first simulation, the GSCLB policy clearly betters the CLB policy, as we would expect. In the third simulation, however, the processing rates are just too skewed along the paths and the CLB policy performs better than the GSCLB policy. In the third simulation, notice that two of the three buffers in every path are serviced at one-quarter of the rate that they are serviced at in the distributed CLB policy. In the second simulation, we attempted to simulate a re-entrant line with "typical" skew along the paths. In this

Table 5.5 Re-entrant Line FMS: Average Backlog

	Policy	
Simulation	GSCLB	CLB
1	1103	1816
2	1036	1431
3	1103	954

particular case, the GSCLB policy was able to perform significantly better than the distributed CLB policy.

Finally, we would like to note that he study of re-entrant lines in [42] considers FMS composed of machines which require no set-up times. Because of this, they study policies such as first-buffer first-serve (FBFS) and last-buffer first-serve (LBFS) which are not clearing policies. Because of the nonclearing nature of these policies, they are not, in general, competitive with clearing policies in controlling FMS with nonzero set-up times. While [42] includes extensive simulation results, we are not able to compare our results with theirs because of the inherent difference in the types of systems considered.

5.5.2 Feedforward Line

Another FMS topology common in industry is the feedforward line. An example of a feedforward line is shown in Figure 5.5. As before, the buffers are labeled according to the convention that we established earlier.

We will perform two simulations of the feedforward line of Figure 5.5 with two different sets of processing rate parameters. For each simulation, we will assume that all buffers have a set-up time of 10.0 and that the network input rates are all 1.0. The buffer processing rate parameters that we will use for each simulation are listed in Table 5.6. Notice that the first simulation has processing rates that are uniform along the paths, the second simulation has processing rates that are quite skewed along the paths.

Once again, we use distributed CLB (supervised by USSM) and GSCLB (supervised by GSPC). The supervisory GSPC policy enforces the service sequence S_1, S_2, S_3, S_4. This GSPC policy can easily be shown to be stable by Theorem 5.8 for each of the two simulation parameter sets. As expected, the GSCLB policy performed significantly better than the CLB policy in the first simulation where there was no processing rate skew along any of the paths. In the second simulation, however, the severe skew of buffer processing rates along each path reduces the performance advantage of the GSCLB policy to almost zero. The magnitude of the processor rate skew is quite severe. In fact, from Table 5.6, we see that in the second simulation, four of the five buffers on each path are constrained by the GSCLB policy to be serviced at one-half

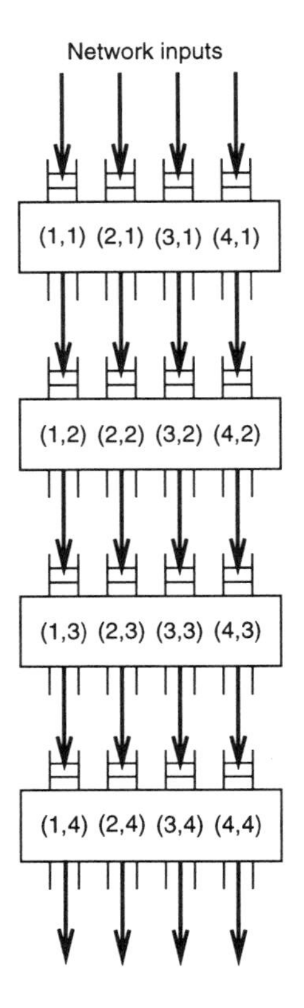

Fig. 5.5 A feedforward line.

of the rate that they are serviced by the distributed CLB policy. The fact that a GSC policy is able to compete with a distributed policy under such unfavorable conditions attests to the advantages of synchronization for feedforward topologies. Table 5.7 summarizes the results of the two simulations.

5.5.3 Cellular Structure I

We now consider the FMS of Figure 5.6, taken from [1]. The purpose of the simulations of this FMS is to demonstrate that the effectiveness of GSC policies versus distributed policies is topology dependent. Notice that the FMS has a highly-interconnected cellular structure. There are only two part types, but because of the interconnected nature of the system, we are forced to define eight paths. Because there are many paths of relatively short length, the advantage gained by synchronization is lessened, and we do not expect GSC policies to perform as well (relative to distributed policies) for this topology as they did for the last two topologies. Clearly, this topology is fundamentally

Table 5.6 Network Input and Buffer Processing Rates

	Simulation	
Buffer	1	2
(1,1)	5.0	5.0
(1,2)	5.0	10.0
(1,3)	5.0	10.0
(1,4)	5.0	10.0
(2,1)	5.0	10.0
(2,2)	5.0	5.0
(2,3)	5.0	10.0
(2,4)	5.0	10.0
(3,1)	5.0	10.0
(3,2)	5.0	10.0
(3,3)	5.0	5.0
(3,4)	5.0	10.0
(4,1)	5.0	10.0
(4,2)	5.0	10.0
(4,3)	5.0	10.0
(4,4)	5.0	5.0

Table 5.7 Feedforward FMS: Average Backlog

	Policy	
Simulation	GSCLB	CLB
1	80.2	110.3
2	80.2	83.0

different than that of either the highly re-entrant structure or the feedforward structure. In this case, each of our S_i will contain more than one path. In particular, we choose $S_1 = \{1, 6, 4\}$, $S_2 = \{2, 5, 8\}$, and $S_3 = \{3, 7\}$.

We will perform three simulations of the cellular structure of Figure 5.6 with three different sets of network input and buffer processing rate parameters. For each simulation, we will assume that all buffers have a set-up time of 5.0. The network input and buffer processing rate parameters that we will use for each simulation are listed in Table 5.8, where in each field the first entry is a network arrival rate (if applicable) and the second entry is a buffer processing rate. Notice that in the first simulation, the processing rates are uniform along the paths. The second and third simulations introduce significant processing rate skew along the individual paths.

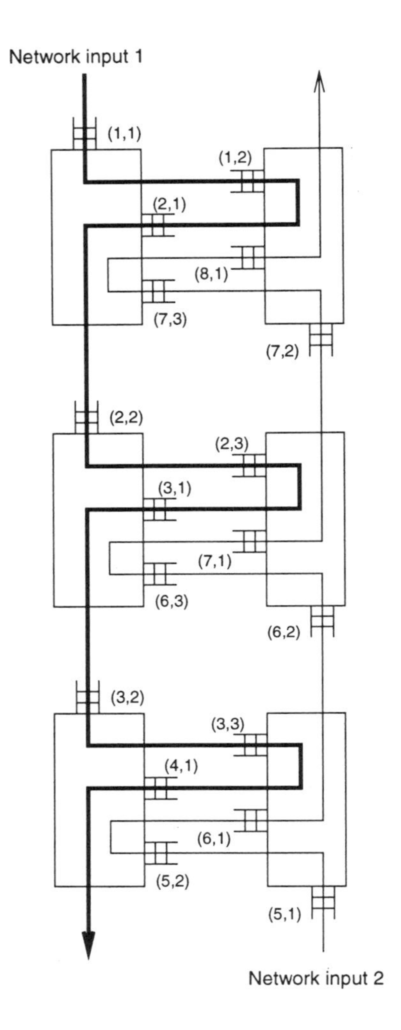

Fig. 5.6 Cellular structure I.

We will use the GSCLB policy (supervised by GSPC) and distributed CLB (supervised by USSM). The CLB policy has consistently performed better than CAF and the "CPK policy" of [86] for this type of topology (for a simulation-based comparison of CLB, CAF, and CPK, see [1]). The supervisory GSPC policy enforces the service period S_1, S_2, S_3. This GSPC policy can be shown to be stable by Theorem 5.9 for each of the three simulation parameter sets.

Table 5.9 summarizes the results of the three simulations. In the first simulation, the GSCLB performs significantly better than does CLB, and surprisingly, in the second simulation GSCLB does even better with respect

Table 5.8 Network Input and Buffer Processing Rates

	Simulation					
Buffer	1		2		3	
(1,1)	0.9	4	1.0	10.0	0.67	2.5
(1,2)	-	4	-	8.0	-	2.5
(2,1)	-	4	-	20.0	-	5.0
(2,2)	-	4	-	20.0	-	5.0
(2,3)	-	4	-	10.0	-	10.0
(3,1)	-	4	-	8.0	-	2.5
(3,2)	-	4	-	20.0	-	3.33
(3,3)	-	4	-	10.0	-	6.67
(4,1)	-	4	-	10.0	-	10.0
(5,1)	1.1	5	2.0	20.0	0.5	2.0
(5,2)	-	5	-	20.0	-	5.0
(6,1)	-	5	-	10.0	-	10.0
(6,2)	-	5	-	8.0	-	5.0
(6,3)	-	5	-	10.0	-	3.33
(7,1)	-	5	-	20.0	-	2.5
(7,2)	-	5	-	10.0	-	3.33
(7,3)	-	5	-	20.0	-	10.0
(8,1)	-	5	-	10.0	-	10.0

to CLB. In the third simulation, however, CLB dramatically outperforms GSCLB. What is the difference between the parameter sets of the second and third simulations that causes such a shift in performance? First of all, the processing rates along the paths are slightly more skewed in the third simulation than in the second. More importantly, however, is the loading of the paths. The load on a particular path is computed by dividing the network input rate of the part type that is processed along the path by the path processing rate. In the third simulation, the three paths with the greatest loads are path 1 at 0.268, path 3 at 0.268, and path 5 at 0.250. Notice that all three of these highest loaded paths are in different processing groups (i.e., $1 \in S_1$, $5 \in S_2$, and $3 \in S_3$). The result of this is that a given less heavily-loaded path that is processed along with one of the heavily loaded paths is either under-utilized (processing occurs at the applicable network input rate, which is less than the path processing rate) or not utilized at all (if the path is not fed by a network input) during a significant portion of the production run. A better strategy is to try to group paths with similar loading together for processing. Unfortunately, for the parameter set of simulation 3, we are unable to find a superior path grouping due to the highly interconnected nature of the topology.

Table 5.9 Cellular Structure I: Average Backlog

	Policy	
Simulation	GSCLB	CLB
1	92.2	136.7
2	53.4	121.9
3	70.2	44.4

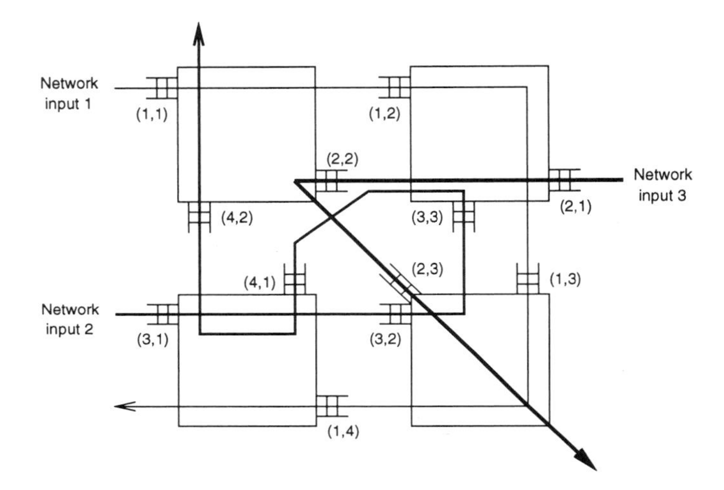

Fig. 5.7 Cellular structure II: path choice 1.

5.5.4 Cellular Structure II

For this set of simulations, we consider the cellular structure FMS of Figure 5.7, taken from [1]. The purpose of these simulations is to illustrate the difference that path selection and grouping can make in performance (an issue that was raised in the last example). We will run two simulations in this section, one with the path choice of Figure 5.7 and one with the path choice of Figure 5.8. For both simulations, we fix the set-up time for all buffers at 10. The network input and processing rates for both simulations are given in Table 5.10 (the buffer indices listed are with respect to Figure 5.7).

In Figure 5.7, we have chosen the paths by making each of them as long as possible (i.e., we allow each path to continue until it reaches either a re-entry point or a network output point). This leads to four paths, which we must group as follows: $S_1 = \{1\}$, $S_2 = \{2\}$, $S_3 = \{3\}$, and $S_4 = \{4\}$. We will once again simulate using the GSCLB (supervised by GSPC) and distributed CLB (supervised by USSM) policies. Distributed CLB has consistently performed better than CAF and the "CPK policy" of [86] for this type of topology (for a

Table 5.10 Network Input and Buffer Processing Rates

Buffer	Input	Processing
(1,1)	0.5	2.5
(1,2)	-	5.0
(1,3)	-	2.5
(1,4)	-	2.0
(2,1)	2.0	6.67
(2,2)	-	20.0
(2,3)	-	20.0
(3,1)	1.0	10.0
(3,2)	-	20.0
(3,3)	-	5.0
(4,1)	-	10.0
(4,2)	-	20.0

Table 5.11 Cellular Structure II: Average Backlog

	Policy	
Simulation	GSCLB	CLB
Path Choice 1	105.6	93.6
Path Choice 2	73.3	93.6

simulation-based comparison of CLB, CAF, and CPK, see [1]). The GSCLB policy is supervised by the GSPC policy that enforces the service period S_1, S_2, S_3, S_4, which can easily be shown to be stable by Theorem 5.8. The results of this simulation are given in the first row of Table 5.11. Notice that the CLB policy performs better than the GSCLB policy. The most obvious reason that the distributed policy performs better than the GSC policy in this case is that whenever GSCLB decides to process S_2, S_3, or S_4, at least one of the four machines is sitting idle.

Hence, we would expect to improve the performance of GSCLB relative to CLB if we could choose our paths so that we could minimize or eliminate machine idling. Also, notice that the two highest loaded buffers are $(2,1)$ and $(1,4)$ (in Figure 5.7). Including buffer $(1,4)$ in path 1 slows down the other three buffers on path 1. Also, as we learned from the simulations in the last section, it may be a good idea to group together paths of similar loading. In an attempt to satisfy all of these intuitive approaches to improve GSCLB performance, we propose the path selections of Figure 5.8. If we group the paths so that $S_1 = \{1, 4\}$, $S_2 = \{2, 3\}$, and $S_3 = \{5\}$, we accomplish all of our goals. As is evident from the second row of Table 5.11, this path selection

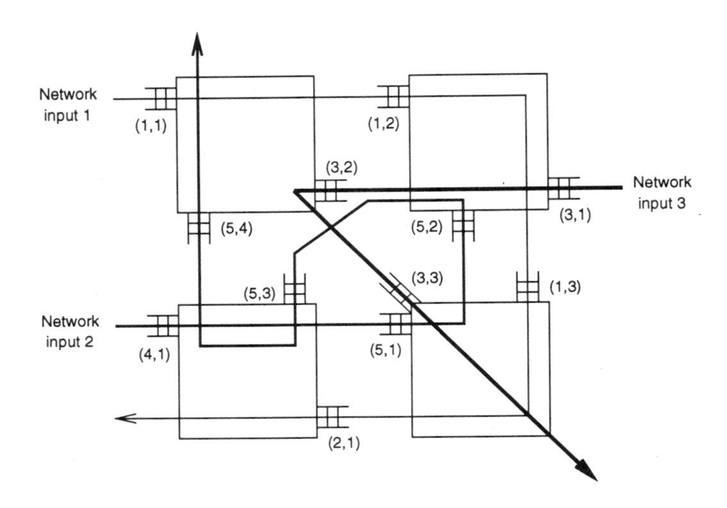

Fig. 5.8 Cellular structure II: path choice 2.

and grouping allows the GSCLB policy to outperform the distributed CLB policy.

5.5.5 Alternatives to GSCLB

Every GSC policy that we have ever simulated eventually falls into a periodic service sequence. This leads us to wonder what is the optimum periodic service sequence for a particular FMS. This is, of course, a very difficult question that we will not attempt to answer. However, what we will demonstrate is that there are alternatives to the GSCLB policy which in some cases perform better than GSCLB.

In particular, we will consider a new GSC policy, which makes its scheduling decisions based on current and past information. Define a vector

$$\chi(p) = [\chi_1(p), \chi_2(p), \cdots, \chi_M(p)]^\top \tag{5.143}$$

such that $\chi_i(p)$ is what the sum of the head buffers of the paths in S_i was at time p', where $t_{p'} < t_p$ is the time that the most recent production run (before time t_p) was begun on the paths in S_i. We can now define a new policy, which we call GSCF. At time t_p, GSCF chooses to process the paths in some S_i, $i \in \{1, 2, \ldots, M\}$, such that

$$\left(\sum_{j \in S_i} x_j(t_p) \right) - \chi_i(p) \geq \left(\sum_{j \in S_q} x_j(t_p) \right) - \chi_q(p) , \tag{5.144}$$

for all $q \in \{1, 2, \ldots, M\}$, $q \neq i$. From this definition, we see that GSCF simply chooses to process the set of buffers whose sum of parts has increased

Table 5.12 Input Rates and Path Processing Rates

Path	Input	Processing
1	10.0	41.0
2	5.0	21.0
3	2.0	9.0
4	1.0	5.0

Table 5.13 Performance of GSCF vs. GSCLB

	GSCLB	GSCF
Ψ	660.1	634.5

the most (or decreased the least) since the last time it was processed. In this way, GSCF will seek a frequency of service for each S_i that will try to minimize the buffer levels (i.e., it will tend to process an S_i more frequently if its buffer levels are rising faster). It is due to this servicing *frequency* idea that we use "F" in the acronym GSCF. Actually, a distributed scheduling policy that operates using principles analogous to GSCF is the DCF policy in [1].

Consider the feedforward line FMS of Figure 5.5 with the network input and path processing rates of Table 5.12. For this FMS, the GSCF policy exhibits better performance than the GSCLB policy, as noted in Table 5.13.

It is interesting to consider the periods that the GSCF and GSCLB policies eventually fall into. The GSCF policy falls into the length-six service period S_1,S_4,S_2,S_1,S_3,S_2, and the GSCLB policy falls into the length-seven service period S_1,S_2,S_1,S_4,S_1,S_2,S_3. If we simulate the FMS with the GSPC policy using the two periods yielded by the GSCLB and the GSCF policies, virtually identical performance measures result.

We can employ Theorem 5.8 to analyze the GSPC policies which result from enforcing the servicing periods yielded by the GSCF and GSCLB policies (because, in this case, $n_i = 1$ for all i). Notice that for the period of GSCF there are six different choices for A and b (from Theorem 5.8) and for GSCLB there are seven different choices for A and b. This is because the matrices A and b are computed differently depending on where the beginning of the service period is taken. If we let A^*_{GSCF} and b^*_{GSCF} be the matrices which yield the largest value of $||(I - A)^{-1}b||$ for the service period of the GSCF policy, we find that

$$||(I - A^*_{GSCF})^{-1}b^*_{GSCF}|| = 2392 . \qquad (5.145)$$

If we similarly define A^*_{GSCLB} and b^*_{GSCLB}, we see that

$$||(I - A^*_{GSCLB})^{-1}b^*_{GSCLB}|| = 2837 . \qquad (5.146)$$

These results are consistent with our simulation, although they do not definitively prove that the GSCF policy will perform better than the GSCLB policy with respect to our performance measure, Ψ. However, they do suggest yet another method for designing periodic service sequences that exhibit good performance.

5.6 SUMMARY

In this chapter we have used scheduling problems in flexible manufacturing systems as a case study in how to analyze boundedness properties of DES. We showed how to prove that several types of local policies will ensure bounded buffer levels. We also showed how to exploit information from the paths on which parts travel to make scheduling more efficient. The main topics covered in this chapter are the following:

- Clear-a-fraction policy.
- Clear-average-oldest-buffer policy.
- Random part selection policy.
- Multiplexer, demultiplexer, and bounded delay elements.
- Stream modifier.
- Global synchronous CAF policy.
- Global synchronous priority policy.
- FMS examples.

Essentially, this is a list of the major topics covered in this chapter.

5.7 FOR FURTHER STUDY

This chapter is based on the work in [16, 17, 14]. In [86], the authors study machines both in isolation and when interconnected in a non-acyclic fashion (i.e., when parts can re-visit the same machine for processing) and analyze the stability of various scheduling policies, including the clear-a-fraction (CAF) policy. In addition, in [86] the authors determine a lower bound on the performance of any stable scheduling policy (in terms of average buffer levels) for an isolated machine. In [21], Chase and Ramadge find a new lower bound on performance is achieved by the introduction of machine "idling". Related work is in [48]. In [48], Lou, Sethi, and Sorger improve on the isolated machine CAF buffer bound of [86]. In [43], Kumar and Seidman study general issues in stability of FMS and introduce a universally stabilizing supervisory mechanism (USSM) that can stabilize any scheduling policy. In [42], Lu and Kumar prove stability for a class of policies whose aims are to optimize the performance of the FMS with respect to criteria such as the mean and variance of the total delay incurred in processing a part (other related work can be found in [97]). In [38], Humes introduces the use of "regulators" (devices which control the rates of part flows throughout an FMS, similar to our stream modifier) to stabilize general nonacyclic FMS, and in [85], Perkins, Humes, and Kumar explain how to generate buffer bounds in a completely regulated FMS (i.e., one that has regulators directly before each machine buffer input). The bounded delay element is essentially the discrete-time version of the bounded delay in [23]. Our stream modifier, multiplexer, and demultiplexer elements are similar to (or, at least, inspired by) the (σ, ρ) regulator, the FCFS MUX, and the DEMUX, respectively, which are detailed in [23]. Certain first-come-first-served (FCFS) networks have been shown to be unstable in [94].

5.8 PROBLEMS

Problem 5.1 (Priority Policy Case Study): This problem is an extended case study in the analysis of boundedness properties of a single machine (in Figure 5.1) which are capable of servicing parts of type i, such that $i \in P$, where $P = \{1, 2, \ldots, N\}$. We fix the rate of arrival of parts to the machine. The machine can only service one part at a time and must be configured differently to service parts of different types. There is a set-up time when reconfiguring the machine for processing different part types. Parts that have arrived at the machine and have not yet been processed are accumulated in buffers. You will show that some such machines can be implemented with buffers of finite size.

While in the chapter we did not use the DES model G to represent the machine, in this problem we will to show you how it is done. Because we are concerned with arrival rates and because the processing of any part takes a finite amount of *real time*, we require that our DES model of the machine be synchronous. The events, e_k, will be required to occur with a fixed real time period. All references to real time will be given in terms of the event period, which we will call a *cycle*. Accordingly, we define the relevant rate and delay constants. There must be $b_i > 0$ cycles between arrivals of parts of type $i \in P$ at buffer i, the machine requires $m_i > 0$ cycles to process one part of type $i \in P$ (when the machine is producing parts of type $i \in P$), and $s_i > 0$ cycles are required to configure the machine to produce parts of type $i \in P$.

We further define

$$w_i = \frac{m_i}{b_i}$$

$$w = \sum_{i \in P} w_i.$$

From the definitions of m_i and b_i, w_i is the number of parts that can arrive at buffer i per every part of type i that enters the machine to be processed (when parts of type i are being processed). In other words, the frequency of arrivals of part type i is 1 part per b_i cycles and the frequency of processing of part type i is 1 part processed per m_i cycles (assuming the machine is currently processing parts of type i); w_i is the ratio of the frequency of arrivals to the frequency of departures. Clearly, it is not possible to bound the buffer levels in general if $w > 1$. Hence, we require that $w < 1$ so that $w_i < 1$ for all $i \in P$.

Let $\mathcal{X} = \Re^N$. The number of parts in buffer $i \in P$ at time $k \in \mathbb{N}$ is x_i, and $\boldsymbol{x}_k = [x_1, x_2, \ldots, x_N]^\top$. Let e_i^b represent that one part of type $i \in P$ arrives at buffer i, let e_i^m represent that one part of type $i \in P$ enters the machine for processing, and let e^0 be the null event. Let $B = \{e_i^b : i \in P\}$ and $M = \{e_i^m : i \in P\}$, so that the set of events is

$$\mathcal{E} = \mathcal{P}\left(B \bigcup M \bigcup \{e^0\}\right) - \{\emptyset\}.$$

Notice that each event $e_k \in \mathcal{E}$ is defined as a set of "subevents".

We now specify g and f_e for $e_k \in g(\boldsymbol{x}_k)$. For $e_k \in g(\boldsymbol{x}_k)$, it is necessary that $e_k \in \mathcal{E}$ and that e_k satisfy the following conditions:

- If $e_i^m \in e_k$, then $x_i > 0$.
- If $e^0 \in e_k$, then $e_k = \{e^0\}$.

If $\boldsymbol{x}_{k+1} = f_{e_k}(\boldsymbol{x}_k)$, then

$$x_i(k+1) = \begin{cases} x_i, & e_i^b \in e_k,\ e_i^m \in e_k \\ x_i + 1, & e_i^b \in e_k,\ e_i^m \notin e_k \\ x_i - 1, & e_i^b \notin e_k,\ e_i^m \in e_k \\ x_i, & e_i^b \notin e_k,\ e_i^m \notin e_k. \end{cases}$$

If $\boldsymbol{x}_{k+1} = f_{e^0}(\boldsymbol{x}_k)$, then

$$\boldsymbol{x}_{k+1} = \boldsymbol{x}_k.$$

Let $\boldsymbol{E}_v = \boldsymbol{E}$ be the set of valid event trajectories. We further specify the set of allowed event trajectories, $\boldsymbol{E}_a$, in order to specify the manner in which the machine chooses which parts to produce and to guarantee the synchronicity of the machine. We specify that the machine observe a *priority-based part servicing policy.* This policy mandates that, once a production run is begun on a particular part type, the run must continue until the buffer of the chosen part type is emptied. Additionally, the parts in buffer $i \in P$ will be serviced before the parts in buffer $i+1$. Once all buffers have been serviced, the first buffer will be serviced again. For $\boldsymbol{E}_a \subset \boldsymbol{E}_v$, every $E \in \boldsymbol{E}_a$ must satisfy the following conditions:

- $e_i^b \in e_{k'}$ for some $0 \leq k' \leq b_i$ for all $i \in P$.
- If $e_i^b \in e_k$, then $e_i^b \in e_{k+b_i}$, and $e_i^b \notin e_{k'}$ for all k', $k < k' < k + b_i$.
- Let $k^* = \min\{k \geq 0 : \ x_i(k) > 0 \ \text{ for some } \ i \in P\}$ and $i^* = \min\{i \in P : x_i(k^*) > 0\}$. $e_{i^*}^m \in e_{k^*+s_{i^*}}$ where (i) $e_i^m \notin e_{k^*+s_{i^*}}$ for all $i \neq i^*$ and (ii) $e_i^m \notin e_{k'}$ for all $i \in P$ and all k', $0 \leq k' < k^* + s_{i^*}$.
- If $e_i^m \in e_k$, then
  - (i) if $x_i(k+1) > 0$, then
    - (a) $e_i^m \in e_{k+m_i}$ and $e_i^m \notin e_{k'}$ for all k', $k < k' < k + m_i$.
    - (b) for all $j \in P$, $j \neq i$, $e_j^m \notin e_{k'}$ for all k', $k < k' \leq k + m_i$.
  - (ii) if $x_i(k+1) = 0$, then
    - (a) $e_j^m \in e_{k+s_j}$ and $e_j^m \notin e_{k'}$ for all k', $k < k' < k + s_j$, where $j = i + 1$ if $(i+1) \in P$ or $j = 1$ if $(i+1) \notin P$.
    - (b) for all k', $k < k' \leq k + s_j$, $e_i^m \notin e_{k'}$ for all $i \in P$, $i \neq j$.
- For any $k' \geq 0$, if $e_i^b \notin e_{k'}$ for all $i \in P$ and $e_i^m \notin e_{k'}$ for all $i \in P$, then $e_{k'} = \{e^0\}$.
- The *real time* between events e_k and e_{k+1} is fixed for all $k \geq 0$.

Notice that with this definition, for start-up the priority-based policy sets up for and processes the first part to arrive that has the highest priority. Following this, it cycles through the processing of part types according to their fixed priority ordering (where after the lowest priority part is processed, the highest priority part is serviced again).

(a) Show that the machine whose operation is described above can be implemented with finite buffers by showing that the machine with a priority-based part processing policy possesses Lagrange stability (i.e., according to Theorem 3.6). In particular, choose

$$\mathcal{X}_b = \{[0, 0, \ldots, 0]^\top\}$$

and

$$\rho(\boldsymbol{x}_k, \mathcal{X}_b) = \sum_{i \in P} m_i x_i. \tag{5.147}$$

For any variable a_i, which is defined for all $i \in P$, let $\underline{a} = \min_i\{a_i\}$ and $\bar{a} = \max_i\{a_i\}$. Prove the following theorem.

Theorem 5.10 (Priority-Based Part Processing Policy): *A machine with a priority-based part servicing policy possesses Lagrange stability w.r.t.* $\boldsymbol{E}_a$ *and* $\mathcal{X}_b$.

(b) Give the actual bounds for the individual buffer levels.

(c) If you increase N, w, or $\sum_{i \in P} m_i x_i(0)$ for the priority-based part servicing policy will you create the possibility that buffer levels can rise even higher, or lower?

(d) If all the arrival rates are the same and all the processing times are the same then in this special case is the priority-based policy a special case of the CAF policy?

Problem 5.2 (Simulation of a Machine): Simulate the machine described in Problem 5.1 under the following conditions (assume that there are three buffers)

(a) With the priority policy.

(b) With the CAF policy.

(c) With the CAOB policy.

Problem 5.3 (Stability Proofs): Repeat the following proofs, adding all the details.

(a) Prove that the CAF policy results in a stable machine.

(b) Prove that the CAOB policy results in a stable machine.

(c) Prove that the RPS policy results in a stable machine (for both sets of constraints).

(d) Prove that the stream modifier policy results in a stable stream modifier.

(e) Prove that the GSCAF policy results in a stable machine.

(f) Prove that the GSPC policy results in a stable machine.

(g) Prove that the GSCLB policy results in a stable machine.

Problem 5.4 (Traffic System Scheduling)*: In this problem you will extend the modeling and analysis results of this chapter to the problem of scheduling vehicle traffic flows on our street/highway system.

(a) What types of intersections can you model? Do you allow traffic in both directions? Do you allow turn lanes? How?

(b) What types of sensors/actuators are needed for implementation?

(c) What constraints will you place on how the intersections are interconnected?

(d) Specify a model G for your traffic intersection.

(e) What overall approach do you want to use for scheduling (path-based or distributed)? Why?

(f) Pick a specific traffic system and simulate it.

(g) Provide a stability analysis to show that the queues in the traffic system remain bounded. Can you also show that the roads do not have to be unreasonably long under heavy traffic conditions?

6

Intelligent Control Systems

6.1 OVERVIEW

Control systems, where the controller is developed using ideas from the functionality of intelligent biological systems, or ones that perform tasks recently only performed by humans, are sometimes called *intelligent control systems.* For instance, when an artificial neural network (a circuit or computer emulation of a biological neural network), fuzzy system (a system designed to model how we reason under uncertainty), or genetic algorithm (a computer algorithm that emulates survival of the fittest and inheritance to simulate evolution) is used as a controller (or in a controller) some will call the resulting control system an *intelligent control system.* In this chapter, we study one class of intelligent control system called an *expert control system.*

An expert system is a computer program that is designed to emulate a human's skills in a specific problem domain. If it is designed to emulate the expertise of a human in performing control activities, it is called an "expert controller." An expert control system (ECS) utilizes an expert controller to interpret plant outputs and reference inputs, to reason about alternative control strategies, and to generate inputs to the plant to improve the performance of the closed-loop system. In this chapter we show how to represent the "rule-based" expert controller (including the inference engine and knowledge-base) and ECS with a mathematical model that is a slightly generalized version of the G model in Chapter 2. Then we show that it is possible to characterize and formally analyze reachability, cyclic behavior, and stability properties for the expert controller and ECS. We overview approaches to verify the qualitative properties of isolated rule-based expert controllers. Finally, we illustrate the results by showing how to perform modeling, analysis, and design of an ECS for a tank liquid level regulation problem and a part balancing problem in flexible manufacturing systems (a much simpler balancing problem than those studied in Chapter 4).

Next, we explain how a planning system (a computer program that emulates how an expert would plan and execute a sequence of actions) can be used as a controller. We discuss the meaning of, for example, stability and reachability properties. Following this, we overview the area of intelligent autonomous control and explain the relevance of the area of hybrid system theory to it. This also serves as a summary of future directions in stability analysis.

While this third case study in stability analysis does show how to study a variety of stability properties, the other chapters do not depend on this one. To understand this chapter you only need to study Chapter 3, since the model in Chapter 2 is extended and presented again here.

6.2 EXPERT CONTROL SYSTEMS

The use of techniques from applied artificial intelligence (AI) has resulted in the construction of extremely complex "rule-based" expert systems. Often such expert systems are being utilized in *critical environments*, where hazards can occur, safety of a crew is an issue, and real-time constraints must be met. For instance, some expert systems for aircraft applications are used for mission planning, while in process control they can be used for diagnosing plant failures. Most often such expert systems are constructed and implemented without any *formal analysis* of the dynamics of how they interface to their environment and how the inference mechanism reasons over the information in the knowledge-base. Currently many expert systems are evaluated either (1) in an empirical manner by comparing the expert system against human experts, (2) by studying reliability and user friendliness, (3) by examining the results of extensive simulations, or (4) by using software engineering approaches.

In this chapter we present a case study that focuses on how to mathematically verify qualitative properties of *general* rule-based expert systems that are used in closed-loop feedback control of dynamical systems.

The expert control system, which consists of the expert controller and plant, is shown in Figure 6.1. In this chapter we focus on the class of expert controllers that has a knowledge-base which consists of *rules* that characterize strategies on how to control the plant. The inference engine is designed to emulate the control expert's decision-making process in collecting reference inputs and plant outputs and reasoning about what command input to generate for the plant. While we use the AI terminology for expert systems, we question the validity of the standard AI models in representing the actual human cognitive structure and processes. The focus here is not on whether we have

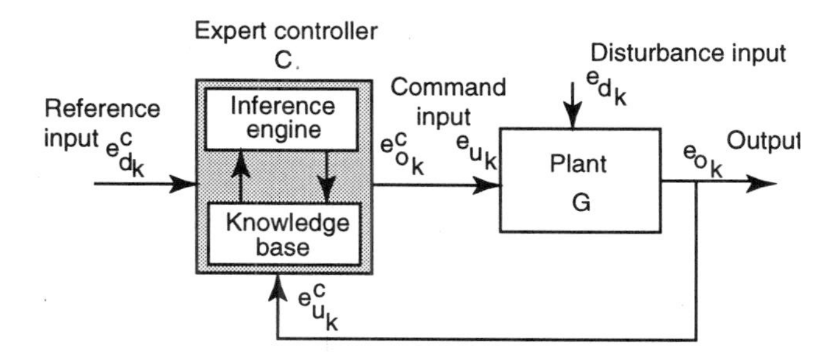

Fig. 6.1 Expert control system.

a good model of the human expert, but rather on whether the heuristic design process for expert controllers produces an ECS that performs adequately. The first step in formally verifying the behavior of the ECS is to develop a mathematical model for the ECS.

Section 6.2.1 shows how mathematical models can be utilized to represent the plant and the expert controller shown in Figure 6.1. After presentation of the model, we discuss how to *design* the expert controller. In particular, we explain issues in the choice of conflict resolution strategies and discuss two general approaches to knowledge-base design: (1) the standard knowledge-acquisition approach to expert system design, and (2) a model-based approach, where the plant model is incorporated into the knowledge-base.

Overall, the expert controller must be designed so that it can coordinate the use of the plant outputs and reference inputs to decide what plant command input to generate so that the closed-loop specifications are met. After the expert controller has been designed and the full closed-loop system model of the ECS has been specified, the dynamic properties of the ECS have to be carefully analyzed to ensure that the system performs within the desired specifications. Mathematical verification of ECS behavior can help ensure that the system will possess certain desirable properties: (1) stability (e.g., to guarantee that the expert controller can keep plant variables bounded), (2) cyclic behavior (e.g., to check that the expert controller will not get stuck in an inappropriate infinite loop where it could, perhaps, exhibit circular reasoning before appropriate conclusions are reached), and (3) reachability (e.g., to guarantee that the expert controller will be able to *infer* appropriate conclusions from certain plant conditions). In this chapter, techniques for reachability analysis and for stability analysis are used for the analysis of both the expert controller and the full ECS.

6.2.1 The Expert Controller

Following a conventional control-theoretic approach, we begin by introducing a mathematical model for the plant (G) shown in Figure 6.1. This will be a slightly more general model than the G model in Section 2.2 in Chapter 2.

The Plant Model It is assumed that the plant can be represented with a model G, where

$$G = (\mathcal{X}, \mathcal{E}, f_e, \delta_e, g, \mathbf{E}_v) \tag{6.1}$$

and

- $\mathcal{X}$ is the set of plant *states* denoted by x,
- $\mathcal{E} = \mathcal{E}_u \cup \mathcal{E}_d \cup \mathcal{E}_o$ is the set of *events*, where
  1. $\mathcal{E}_u$ is the set of *command input events* of the plant,
  2. $\mathcal{E}_d$ is the set of *disturbance input events* of the plant,
  3. $\mathcal{E}_o$ is the set of *output events* of the plant,
- $g : \mathcal{X} \rightarrow \mathcal{P}(\mathcal{E}_u \cup \mathcal{E}_d) - \{\emptyset\}$, where $\mathcal{P}$ denotes the power set, is the *enable function*,
- $f_e : \mathcal{X} \rightarrow \mathcal{X}$ for $e \in \mathcal{P}(\mathcal{E}_u \cup \mathcal{E}_d) - \{\emptyset\}$ are the *state transition maps*,
- $\delta_e : \mathcal{X} \rightarrow \mathcal{E}_o$ for $e \in \mathcal{P}(\mathcal{E}_u \cup \mathcal{E}_d) - \{\emptyset\}$ are the *output maps*,
- $\mathbf{E}_v$ is the set of all *valid event trajectories* (event trajectories that are *physically* possible).

When discussing the states and events at time k, $x_k \in \mathcal{X}$ is the plant state, $e_{u_k} \in \mathcal{E}_u$ is a command input event of the plant, $e_{d_k} \in \mathcal{E}_d$ is a disturbance input event of the plant, and $e_{o_k} \in \mathcal{E}_o$ is an output event of the plant. We will call each $e_k \subset g(x_k)$ an *event* that is *enabled* at time k. Note that e_k is a *set* of the command and disturbance input events described above. For convenience we assume that e_k contains at most one event of each type $\mathcal{E}_u$ and $\mathcal{E}_d$. If an event is enabled then it is possible that it *occurs*. If an event e_k occurs at time k and the current state is x_k, then the next state is $x_{k+1} = f_{e_k}(x_k)$ and the output is $e_{o_k} = \delta_{e_k}(x_k)$. We require that f_e and δ_e be defined when $e \subset g(x)$ and $\mathbf{E}_v$ allows e to occur (as will be explained next).

Any sequence $\{x_k\}$ (finite or infinite in length) such that for all k, $x_{k+1} = f_{e_k}(x_k)$, where $e_k \subset g(x_k)$ is called a *state trajectory*. The set of all *event trajectories* is composed of those sequences $\{e_k\}$, such that there exists a state trajectory $\{x_k\}$, where for all k, $e_k \subset g(x_k)$. Hence to each event trajectory, which specifies the order of the application of the function f_e (order that the events fire) there corresponds a unique state trajectory (but, in general, not vice versa). Define $\mathbf{E}$ to be the set of all infinite and finite length event *trajectories* (sequences of events $e \in \mathcal{P}(\mathcal{E}_u \cup \mathcal{E}_d) - \{\emptyset\}$) that can be generated by g and f_e. The set $\mathbf{E}_v \subset \mathbf{E}$ is the set of all physically possible event trajectories; hence, even if $x_k \in \mathcal{X}$ and $e_k \subset g(x_k)$, it is not the case that e_k can occur unless it lies on a valid event trajectory that ends at x_{k+1}, where $x_{k+1} = f_{e_k}(x_k)$.

Model of the Expert Controller The expert controller shown in Figure 6.1 has two inputs; the reference input events $e_d^c \in \mathcal{E}_d^c$ and the output events of the plant $e_o \in \mathcal{E}_o$. Based on its state and these inputs, the expert controller generates enabled command input events to the plant $e_o^c \in \mathcal{E}_o^c$. Hence, the expert controller models how a "human in-the-loop" coordinates the use of (1) feedback information from the plant (2) reference inputs (modeling the current control objectives), and (3) information in its own memory (the controller state). Moreover, the expert controller models the cognitive processes used to decide how to use all this information to decide what control actions to execute. We will often speak of the interactions between the inference engine and knowledge-base shown in the expert controller in Figure 6.1 as forming an "inference loop." This inference loop constitutes the core of the expert controller where information in the knowledge-base is interpreted by the inference engine, actions are taken, the knowledge-base is updated, and the process repeats (i.e., "loops").

The full expert controller shown in Figure 6.1 is modeled by C, where

$$C = (\mathcal{X}^c, \mathcal{E}^c, f_e^c, \delta^c, g^c, \mathbf{E}_v^c), \tag{6.2}$$

where

- $\mathcal{X}^c = \mathcal{X}^b \times \mathcal{X}^i$ is a set of *expert controller states* x^c, where $\mathcal{X}^b$ is the set of knowledge-base states $\mathbf{x}^b$ and $\mathcal{X}^i$ is the set of inference engine states $\mathbf{x}^i$ to be defined below,
- $\mathcal{E}^c = \mathcal{E}_u^\ell \cup \mathcal{R} \cup \mathcal{E}_o^c$ is the set of *events* of the expert controller, where
  1. $\mathcal{E}_u^\ell \subset \mathcal{P}(\mathcal{E}_o \cup \mathcal{E}_d^c) - \{\emptyset\}$ is the set of sets of *reference input* ($\mathcal{E}_d^c$) and *plant output events* ($\mathcal{E}_o$) that can occur for which the expert controller will have to know how to respond.
  2. $\mathcal{R}$ is the set of *rules* in the *knowledge-base* of the expert controller.
  3. $\mathcal{E}_o^c \subset \mathcal{P}(\mathcal{E}_u) - \{\emptyset\}$ is a set of output events of the expert controller (sets of enabled command input events to the plant).
- $g^c : \mathcal{X}^b \times \mathcal{X}^i \to \mathcal{P}(\mathcal{E}_u^\ell \cup \mathcal{R}) - \{\emptyset\}$ is the *enable function*,
- $f_e^c : \mathcal{X}^b \times \mathcal{X}^i \to \mathcal{X}^b \times \mathcal{X}^i$ for $e \in \mathcal{P}(\mathcal{E}_u^\ell \cup \mathcal{R}) - \{\emptyset\}$ are the state transition maps,
- $\delta^c : \mathcal{X}^b \times \mathcal{X}^i \to \mathcal{E}_o^c$ are the output maps (notice the difference from the plant model),
- $\mathbf{E}_v^c \subset \mathbf{E}^c$ is the set of *valid inference loop (expert controller) trajectories* (controller event trajectories that are physically possible)

In this framework, it is assumed that an occurrence of a command input event to the expert controller $e_u^\ell \in \mathcal{E}_u^\ell$ is always accompanied by a firing of an enabled rule $r \in \mathcal{R}$, so that the inference loop can be updated accordingly.

Similarly, a rule $r \in \mathcal{R}$ cannot fire alone, since the inference loop is updated only if there is a change in the plant reflected via its output or a change in the reference input event. Therefore, the expert controller input events must contain exactly one rule $r \in \mathcal{R}$ and one command input event $e_u^\ell \in \mathcal{E}_u^\ell$ and each e_u^ℓ has at most one plant output event $e_o \in \mathcal{E}_o$ and reference input event $e_d^c \in \mathcal{E}_d^c$ contained in it (this can be modeled using $\mathbf{E}_v^c$). The $f_e^c(x^c)$ for $e \subset g^c(x^c)$ are operators describing updates to the knowledge-base and inference engine states when the plant output and/or reference input change and the rule r fires. The output events of the expert controller $\delta^c(x^c)$ are defined to be the enabled command inputs to the plant G. The valid inference loop trajectories $\mathbf{E}_v^c \subset \mathbf{E}^c$ can also put constraints on the input event trajectories based on the rule-base and conflict resolution strategy. Note that the enabled command input events of the plant cannot occur simultaneously as the controller changes state; however, they can occur any time later. The controller can control the enabling of the command input events of the plant; however, it does not have any capabilities to control the plant's disturbance input events. The full specification of C is achieved by defining the rule-base and inference engine for the expert controller, that is, by defining the components of the inference loop.

Modeling a Rule-Base: It is important to note that although the focus in this chapter is on rule-based systems, we are not restricted to modeling only rule-based systems; other AI knowledge representation formalisms can also easily be represented. To see this first note that any system that can be represented with the general, extended, or high-level Petri Net can be represented with C. Then the Petri Net can be used to represent, for instance, *semantic nets, frames,* or *scripts.* Alternatively, one could directly model such knowledge representation schemes with C. Next, we model the rule-base.

Let $A = \{a_1, a_2, ..., a_n\}$ be a set of *facts* that can be true or false (and their truth values can change over time). Let

$$T : A \to \{0, 1\} \tag{6.3}$$

where $T(a_i) = 1 (= 0)$ indicates that a_i is true (false). Let $\Re$ denote the real numbers, $\boldsymbol{V} \subset \Re^m$, and $\boldsymbol{v} \in \mathbf{V}$ denote an m-dimensional column vector of *variables.* We are thinking here of facts and variables in "working memory" [10]. Let $\mathcal{X}^b = \Re^{m+n}$ where $\boldsymbol{x}^b \in \mathcal{X}^b$,

$$\boldsymbol{x}^b = [\boldsymbol{v}^\top, T(a_1), T(a_2), ..., T(a_n)]^\top = [x_1^b, x_2^b, ..., x_{m+n}^b]^\top$$

and let $x_{i_k}^b$ denote the *ith* component of $\boldsymbol{x}^b$ at time k. Let $P_i, i = 1, 2, ..., p$ denote a set of p *premise functions,* that is,

$$P_i : \mathcal{X}^b \times \mathcal{E}_u^\ell \to \{0, 1\} \tag{6.4}$$

and $P_i(\boldsymbol{x}_k^b, e_{u_k}^\ell) = 1 (= 0)$ indicates that $P_i(\boldsymbol{x}_k^b, e_{u_k}^\ell)$ is true (false) at time k. The P_i will be used in the premises of the rules to state the conditions under which a rule is *enabled* (i.e., they model the left-hand sides of rules). Let the *antecedent formulas*, denoted by Φ, be defined in the following recursive manner:

1. $T(a)$ for all $a \in A$, and $P_i, i = 1, 2, \ldots, p$ are antecedent formulas.

2. If Φ and Φ' are antecedent formulas then so are $\neg\Phi, \Phi \wedge \Phi', \Phi \vee \Phi'$, and $\Phi \Rightarrow \Phi'$, (where $\neg$ (not), $\wedge$ (and), $\vee$ (or), $\Rightarrow$ (implies) are the standard Boolean connectives).

3. Nothing else is an antecedent formula unless it is obtained via finitely many applications of 1–2 above.

For example, if $m = 3$, $n = 2$, $A = \{a_1, a_2\}$, $\mathbf{V} \subset \Re^3$, and P_1 tests "$x^b_{2_k} < 5.23$", P_2 tests "$x^b_{3_k} = 1.89$", $e^c_{d_k}$ and e_{o_k} are real numbers, and P_3 tests "$(e^c_{d_k} < 5) \vee (e_{o_k} \geq 2)$", then $\Phi' = P_1 \wedge P_2 \wedge P_3 \wedge (T(a_1) \vee \neg T(a_2))$ is a valid antecedent formula (where $<, \geq$ and $=$ take on their standard meaning).

Let $C_i, i = 1, 2, ..., q$ denote the set of q *consequent functions*, where

$$C_i : \mathcal{X}^b \times \mathcal{E}^\ell_u \rightarrow \mathcal{X}^b \tag{6.5}$$

will be used in the representation of the consequents of the rules (the right-hand sides of the rules), that is, to represent what actions are taken to the knowledge-base when a rule is *fired.* Let the *consequent formulas*, denoted with Ψ, be defined in the following recursive manner:

1. For any $C_i, i = 1, 2, \ldots, q, C_i$ is a consequent formula.

2. For any C_i, C_j, $C_i \wedge C_j$ is a consequent formula.

3. Nothing else is a consequent formula unless it is obtained via finitely many applications of 1–2 above.

Following the above example for the premise formula, C_1 may be $x^b_{4_{k+1}} = T(a_1) := 1$ (make a_1 true), C_2 may mean let $x^b_{2_{k+1}} := x^b_{2_k} + 2.9$, C_3 may mean let $x^b_{3_{k+1}} := e^c_{d_k}/2$, and $\Psi' = C_1 \wedge C_2 \wedge C_3$ makes a_1 true ($x^b_{4_{k+1}} := 1$), increments x^b_2 (variable v_2) and assigns $e^c_{d_k}/2$ to $x^b_{3_{k+1}}$.

Notice that we could also define the C_i, such that $C_i : \mathcal{X}^b \times \mathcal{E}^\ell_u \rightarrow \mathcal{X}^b \times \mathcal{E}^c_o$ but, in this case, care must be taken to ensure that the closed-loop system is properly defined. Moreover, one could define $C_i : \mathcal{X}^b \times \mathcal{X}^i \times \mathcal{E}^\ell_u \rightarrow \mathcal{X}^b \times \mathcal{X}^i \times \mathcal{E}^c_o$, so that the rules could characterize changes made to the inference strategy based on the state of the knowledge-base and/or the reference input (i.e., the inference strategy could be changed based on the current objectives stated in the reference input). Similar, more general definitions could be made for the P_i above. In this chapter we will not consider such possibilities and, hence, we will focus solely on the use of the P_i and C_i defined in Equations (6.4) and (6.5), above.

The rules in the knowledge-base $r \in \mathcal{R}$ are given in the form of

$$r = \text{IF} \quad \Phi \quad \text{THEN}, \quad \Psi \tag{6.6}$$

where the action Ψ can be taken only if Φ evaluates to true. Formally for (6.6),

$$e_k = \{r, e_u^\ell\} \subset g^c(x_k^c)$$

can possibly occur only if Φ evaluates to true at time k for the given state $\boldsymbol{x}_k^b$ and the command input event $e_{u_k}^\ell$. If $e_k \subset g^c(x_k^c)$ occurs, then the next state $x_{k+1}^c = f_{e_k}^c(x_k^c)$ is given by: (1) the application of Ψ to the state $\mathbf{x}_k^b \in \mathcal{X}^b$ to produce $\boldsymbol{x}_{k+1}^b$, and (2) updating the inference engine state $\boldsymbol{x}^i \in \mathcal{X}^i$, which will be discussed below.

For instance, following the above examples, $r' =$IF Φ' THEN Ψ' is a valid knowledge-base rule which is enabled at time k if Φ' evaluates to true at time k for the knowledge-base state $\boldsymbol{x}_k^b$ and the command input $e_{u_k}^{\ell'}$. If $e_k' = \{r', e_{u_k}^{\ell'}\}$ is enabled (which depends on both $\boldsymbol{x}^b$ and $\boldsymbol{x}^i$) and it occurs, the knowledge-base component of the next state of the controller ($\boldsymbol{x}_{k+1}^b$) is produced by applying Ψ' to $\boldsymbol{x}^b$ at time k. The inference engine state is updated based on all enabled rules $r \in \mathcal{R}$ and the fired rule r' (the exact update process is explained below). The inclusion of input events $\mathcal{E}_u^\ell$ in the rule-base allows the expert controller designer to incorporate the plant output feedback and the reference input *variables* directly as parts of the rules. This is analogous to the use of *variables* in conventional rule-based expert systems.

Modeling the Inference Engine: To model the inference mechanism one must be able to represent its three general functional components:

1. *Match Phase.* The premises of the rules are matched to the current facts and data stored in the knowledge-base, and the reference input and plant output.
2. *Select Phase.* One rule is selected to be fired.
3. *Act Phase.* The actions indicated in the consequents of the fired rule are taken on the knowledge-base, the inference engine state is updated, and subsequently the input to the plant is generated.

Here, the characteristics of the "match phase" of the inference mechanism are inherently represented in the knowledge-base. In AI terminology

$$\Gamma_k = \{r : \{r, e_{u_k}^\ell\} \subset g^c(x_k^c) \text{ so that the } \Phi \text{ of rule } r \in \mathcal{R} \text{ evaluates to true for } e_{u_k}^\ell\} \tag{6.7}$$

is actually the knowledge-base "conflict set" at time k (the set of enabled rules in terms of the knowledge-base only). The select phase (which picks one rule from Γ_k to fire) is composed of "conflict resolution strategies" (heuristic inference strategies), of which a few representative ones are listed below:

1. *Refraction.* All rules in the conflict set that were fired in the past are removed from the conflict set. However, if firing a rule affects the matching data of the other rules' antecedents, those rules are allowed to be considered in the conflict resolution.

2. *Recency.* Use an assignment of priority to fire rules based on the "age" of the information in the knowledge-base that matches the premise of each rule. The "age" of the data that matches the premise of a rule is defined as the number of rule firings since the last firing of the rule which allows it to be considered in the conflict set.

3. *Distinctiveness.* Fire the rule that matches the most (or most important) data in the rule-base (many different types of distinctiveness measures are used in expert systems). Here, we will count the number of different premise functions P_i used in the antecedent of a rule and use this as a measure of distinctiveness.

4. *Priority Schemes.* Assign a priority ranking of the rules then choose from the conflict set the highest priority rule to fire.

5. *Arbitrary.* Pick a rule from the conflict set to fire at random.

It is understood that the distinctiveness conflict resolution strategy is actually a special case of a priority scheme but we include both since distinctiveness has, in the past, been found to be useful in the development of expert systems. Note that in a particular expert system any number of the above conflict resolution strategies (in any fixed, or perhaps variable order) may be used to determine which rule from the conflict set is to be fired. Normally, these conflict resolution strategies are used to "prune" the size of the knowledge-base conflict set Γ_k until a smaller set of enabled rules is obtained. These rules are the "enabled rules" in the model C of the combined knowledge-base and inference engine after the conflict resolution pruning. If all the conflict resolution strategies are applied and more than one rule remains, then (5) above ("arbitrary") is applied to randomly fire one of the remaining rules. The act phase will be modeled by the operators f_e^c, which represent the actions taken on the knowledge-base and inference engine if a rule with the corresponding input event to the inference loop occurs.

The priority and distinctiveness of a rule in the knowledge-base are fixed for all time, but the refraction and recency vary with time. Thus, the inference engine state $\boldsymbol{x}^i$ has to carry the information regarding both refraction and recency. Assume that the knowledge-base has n_r rules and the rules are numbered from 1 to n_r. Define a function $\Pi(i)$ to be 1 if rule i is deleted from the conflict set, and 0 if rule i is allowed to be considered in conflict resolution. This function is used for implementing the refraction component of the select phase. Let

$$\boldsymbol{p} = [\Pi(1), \Pi(2), \Pi(3), \ldots, \Pi(n_r)]^\top$$

be an n_r-vector whose components represent whether a rule can be included in the conflict set when it is enabled in state $\mathbf{x}^b$. Let the n_r-vector

$$\boldsymbol{s} = [s_1, s_2, s_3, \ldots, s_{n_r}]^\top$$

where s_i is an integer representing the *age* of information in the knowledge-base, which matches the premise of rule i (to be fully defined below). We will use $\mathbf{s}$ to help represent the recency conflict resolution strategy. The inference engine state is defined as $\boldsymbol{x}^i = [\boldsymbol{p}^\top, \boldsymbol{s}^\top]^\top \in \mathcal{X}^i$.

To complete the model of the expert controller we need to fully define g^c and f_e^c. The state transitions that occur to update $\boldsymbol{p}$ and $\boldsymbol{s}$ are based on the refraction and recency of the information represented by the components of $\boldsymbol{x}^i$. A matrix $\boldsymbol{A}$ is used to specify how to update $\boldsymbol{p}$ and $\boldsymbol{s}$ and is defined to have a dimension of $n_r \times n_r$ and its ij^{th} component, $a_{ij} = 1(0)$ if firing rule i (does not) affects the matching data of rule j. Essentially, $\boldsymbol{A}$ contains static information about the interconnecting structure of the knowledge-base which is automatically specified once the rules are loaded into the knowledge-base and before the dynamic inference process is started. It provides a convenient way to model the recency and refraction schemes.

We use variables $\tilde{e}_i, d_i$, and p_i, for $i, 1 \leq i \leq n_r$ to define the update process for $\mathbf{x}^b, \mathbf{p}$, and $\mathbf{s}$, where $\tilde{e}_i = 1(0)$ indicates that rule i is enabled(disabled), d_i holds the distinctiveness level of rule i (the higher the value is, the more distinctive the rule is), and p_i holds the priority level of rule i (the priority is proportional to the p_i value). The d_i and p_i components are specified when the knowledge-base is defined and they remain fixed. The values of $s_i, \tilde{e}_i$, and $\Pi(i)$ change with time k, so we use $s_i^k, \tilde{e}_i^k$, and $\Pi_k(i)$, respectively, to denote their values at time k.

The inference loop in the expert controller can be executed in the following manner. First, through "knowledge acquisition" the knowledge-base is defined; then $\mathbf{p}, \mathbf{s}$, and $\tilde{e}_i, 1 \leq i \leq n_r$ are initialized to 0. The inference step from k to $k+1$ is obtained by executing the three following steps (we list this in a "psuedocode" form to help clarify how we have done our analysis and simulation for our applications in Sections 5 and 6):

1. *Match Phase*

   FOR rule $r = 1$ TO rule $r = n_r$ DO:
       IF $r \in \Gamma_k$ THEN $\tilde{e}_r^k := 1$ { Finds the enabled rules }
   IF there is just one r' such that $\tilde{e}_{r'}^k = 1$ THEN GOTO the *Act Phase*
   IF there are no r' such that $\tilde{e}_{r'}^k = 1$ THEN STOP { expert controller not properly defined, i.e., it cannot properly react to all possible plant output/reference input conditions. }

2. *Select Phase*

   FOR rule $r = 1$ TO rule $r = n_r$ DO: { Pruning based on refraction }
       IF $\tilde{e}_r^k = 1$ THEN
           IF $\Pi_k(r) = 1$ THEN $\tilde{e}_r^k := 0$
   IF there is just one r' such that $\tilde{e}_{r'}^k = 1$ THEN GOTO the *Act Phase*

IF there are no r' such that $\tilde{e}^k_{r'} := 1$ THEN STOP { Expert controller not properly defined }
LET $s = -\infty$ {Pruning based on recency }
FOR $j = 1$ TO 2 DO: { Search for rule(s) with the lowest age value(s) }
FOR rule $r = 1$ TO rule $r = n_r$ DO:
IF $\tilde{e}^k_r = 1$ THEN
IF $-s^k_r < s$ THEN $\tilde{e}^k_r := 0$
ELSE s:=$-s^k_r$
IF there is just one r' such that $e^k_{r'} = 1$ THEN GOTO the *Act Phase*
LET $d = 0$ {Pruning based on distinctiveness }
FOR $j = 1$ TO 2 DO: { Search for rule(s) with the highest distinctiveness value(s) }
FOR rule $r = 1$ TO rule $r = n_r$ DO:
IF $\tilde{e}^k_r = 1$ THEN
IF $d_r < d$ THEN $\tilde{e}^k_r := 0$
ELSE d:=d_r
IF there is just one r' such that $\tilde{e}^k_{r'} = 1$ THEN GOTO the *Act Phase*
LET $p = 0$ {Pruning based on priority }
FOR $j = 1$ TO 2 DO: { Search for rule(s) with the highest priority }
FOR rule $r = 1$ TO rule $r = n_r$ DO:
IF $\tilde{e}^k_r = 1$ THEN
IF $p^k_r < p$ THEN $\tilde{e}^k_r := 0$
ELSE p:=p_r
LET r' be any r such that $\tilde{e}^k_{r'} = 1$ {Pruning based on "arbitrary" }

3. *Act Phase*

Let $e' = \{r', e^\ell_{u_k}\}$
Let $(\boldsymbol{x}^b_{k+1}, \boldsymbol{x}^i_{k+1}) = f^c_{e'}(x^c_k)$ {Update the knowledge-base state; the state $\boldsymbol{x}^i_{k+1}$ is defined below}
$\Pi_{k+1}(r') := 1$ {Remove rule r' from the conflict set based on refraction}
FOR rule $r = 1$ to rule $r = n_r$ DO
IF $r \in \Gamma_k$ THEN $s^{k+1}_r := s^k_r + 1$ {Increment the matching age for all rules that were in the conflict set (for recency)}
FOR r=1 TO r=n_r DO
IF $a_{r'r} = 1$ THEN $\Pi_{k+1}(r) := 0$ and $s^{k+1}_r := 0$ {Allow the rules affected by the firing of rule r' to be considered in the conflict set and reset ages of these rules to 0}

In the step "pruning based on refraction" where it says "STOP" one could change this to "Reset the $\tilde{e}^k_r$ values to the values they had before entering pruning based on refraction and continue" so that the expert controller uses the refraction conflict resolution strategy only if it reduces the size of the conflict set. Note that $f^c_{e'}(x^c_k)$, where $e' = \{r', e^\ell_{u_k}\}$ is the action defined by the consequent formula of rule r' taken on the current knowledge-base state $\boldsymbol{x}^b_k$ *and* the action defined for updating the inference engine state $\boldsymbol{x}^i_k$. In the steps discussed above, the conflict resolution is done based on refraction, recency, and distinctiveness followed by priority (with "arbitrary" making any final decisions if there is more than one rule). In other cases, the conflict-resolution strategies may have a different order (the choice of the order being dictated by the application at hand). Note that we have not specified exactly how to define the output map δ^c which defines how the input to the plant is generated.

To summarize, the operation of the expert controller proceeds by:

1. Acquisition of $e^{\ell}_{u_k}$, the plant output and reference input events at time k.

2. Forming the conflict set Γ_k in the match phase from the set of rules in the knowledge-base and based on $e^{\ell}_{u_k}$, the current status of the truth of various facts, and the current values of variables in the knowledge-base.

3. The use of conflict resolution strategies (refraction, recency, distinctiveness, priority, and arbitrary) in the select phase to find one rule $r' \in \Gamma_k$ to fire.

4. Executing the actions characterized by the consequent of rule r' in the act phase (this involves updating the knowledge-base and inference engine state and generating the plant input).

The timing of the event occurrences in the expert controller is such that the controller is synchronous with the plant (i.e., if an event occurs in the plant it will cause a rule to fire) and with the reference input (i.e., if a reference input event occurs, the controller will immediately react to it also). Hence, in response to plant output and reference input events, the expert controller generates plant inputs (sets of enabled events).

Expert Controller Design

Conflict Resolution Strategy Design: Often the inference mechanism is specified a priori and independent of the knowledge-base. In general, one would often use the same inference mechanism for a wide variety of plants and just change the knowledge-base to reflect how to appropriately control the particular plant being considered. Alternatively, one could design the inference mechanism of the expert controller. In particular, one can select the order and type of conflict resolution strategies. For instance, for the applications we study later we will omit the use of some of the conflict resolution strategies if they do not truly emulate the proper way to make decisions about the control problem at hand.

One can also modify the conflict resolution strategies presented in the previous subsection. For instance, the conflict resolution strategies can be modified such that every time a rule fires, it allows certain rules to be reconsidered in conflict resolution independent of whether the fired rule affects the matching data of those rules. This type of conflict resolution strategy can be implemented using the same mechanism as discussed above; however, the matrix $\boldsymbol{A}$ must be redefined such that the a_{ij} is $1(0)$ if firing rule i (does not) allows rule j to be considered in the conflict resolution. This conflict resolution strategy may need less memory and be more efficient computationally. However, it is harder to develop since it requires the designer to adjust the matrix $\boldsymbol{A}$ using ad hoc methods until it satisfies the design objectives. Thus, it is suggested to use some standard set of conflict resolution strategies (or a subset of them) and adjust the knowledge-base appropriately. There may be good reasons to omit the use of a particular conflict resolution strategy (besides the fact that it does not properly emulate expert decision-making). For instance, in some cases one may design an inference engine that implements only the distinctiveness, priority and arbitrary conflict resolution strategies. This inference engine is less computationally demanding and easier to implement since the inference engine does not have a state ($\mathcal{X}^c = \mathcal{X}^b$).

Knowledge Acquisition for Control: This subsection describes how the knowledge we have about how to control the plant can be loaded into the knowledge-base. Two versions of knowledge-base design are discussed in this subsection. The first uses a standard approach to expert system design. In the second knowledge-base design approach one incorporates a plant model (or some version of it) as a part of the knowledge-base (i.e., it uses *model-based control* to aid in making control decisions).

For the expert system approach to knowledge-base design one implements the control knowledge via rules that directly relate the reference input and plant output (in the left-hand side of the rule) to the plant input (in the right-hand side of the rule). First, we have to designate a state $\mathbf{x}^b$ which has m variable elements to correspond to the m conditions of the observed plant output and reference input (and possibly other internal variables). These conditions may represent the output, output's rate of change, the reference input, and so on. Next, we have to specify the set of rules r that govern the updating process of the knowledge-base state $\boldsymbol{x}^b$ that depends on the reference input and plant output.

It is important to note here that the consequent formulas of the rules represent how the state in the knowledge-base (information) changes based on the occurrence of input events. However, they do not directly provide the control action to the plant. The control action to the plant is defined via a mapping of the state of the expert controller to the enabled command input events to the plant. However, we can reserve an element in the state $\mathbf{x}^b$ to represent the enabled command input events to the plant. In other words, we can associate an element in the knowledge-base state x_i^b to be the current output of the controller so that $x_i^b \in \mathcal{E}_o^c$. For example, let all controller output events be numbered ($j = 1, 2, \ldots, n$). Then $\delta^c(x_k^c)$ can be defined to be, say $x_{5_k}^b$ (the fifth element of the state $\boldsymbol{x}^b$); then $x_{5_k}^b = j$ means that the enabled command input to the plant at time k is the command input event j.

For some plants, we want to design a controller which maps the output of the plant and the reference input directly to a control action. This can be done in a similar manner to how we did it for the fuzzy controller or by using rules which have a form of "IF P_i THEN C_i" where P_i and C_i are functions of the command input to the inference loop only. For these cases, the state $\boldsymbol{x}^b$ has a dimension of 1, $\boldsymbol{x}^b$ represents the controller output events, and $\delta^c(x^c) = \boldsymbol{x}^b$ (one must be careful however, in making sure that the closed-loop system is properly defined).

Another design approach is to incorporate the plant model in the knowledge-base. This is analogous to model-based approaches used in conventional control systems. The state of the plant can be included as a part in the knowledge-base state $\mathbf{x}^b$. The state of the plant in the knowledge-base must be updated using the output of the plant which may not provide all information regarding the state of the plant itself. Hence, one must be careful to specify a plant model in the knowledge-base that can accurately reflect the dynamical behavior of the plant using available information.

Besides the plant model states, the state $\boldsymbol{x}^b$ may contain the facts truth values $T(a_i)$ corresponding to the history of the plant and controller behavior (past conditions of the plant and the controller), and other variables. Similar to the previous section, the set of knowledge-base rules $r \in \mathcal{R}$ must be specified to represent the state transition function of the knowledge-base state $\boldsymbol{x}^b$ and the enabled command input to the plant. We can also reserve an element in the state $\boldsymbol{x}^b$ to represent the enabled command input events to the plant as in the above design approach. The consequent formulas for this type of model-based design tend to be more complex than the previous ones, since they also update the plant model used in the controller at the same time.

6.2.2 The Expert Control System: Model and Approaches to Analysis

In this Section we define the full closed-loop ECS model, highlight some of the timing characteristics of the ECS, and outline some analysis techniques for the ECS. Then, we prove some general properties of the expert controller and ECS.

The Closed-Loop ECS Model The closed-loop system (ECS), denoted by S, is given by

$$S = (\mathcal{X}^s, \mathcal{E}^s, f_e^s, \delta_e^s, g^s, x_o^s, \mathbf{E}_v^s), \tag{6.8}$$

where

- $\mathcal{X}^s = \mathcal{X} \times \mathcal{X}^c$ is the set of *closed-loop system states* x^s.
- $\mathcal{E}^s = \mathcal{E}_u \cup \mathcal{E}_d^s \cup \mathcal{E}_o$ where $\mathcal{E}_d^s = \mathcal{E}_d \cup \mathcal{E}_d^c$, $\mathcal{E}_d$ and $\mathcal{E}_d^c$ are the *input events* to the closed-loop system, $\mathcal{E}_o$ contains the *plant output events.*
- $g^s : \mathcal{X} \times \mathcal{X}^c \to \mathcal{P}(\mathcal{E}_u \cup \mathcal{E}_d^s) - \{\emptyset\}$ is the *closed-loop enable function.*
- $f_e^s : \mathcal{X} \times \mathcal{X}^c \to \mathcal{X} \times \mathcal{X}^c$ for $e \in \mathcal{P}(\mathcal{E}_u \cup \mathcal{E}_d^s) - \{\emptyset\}$ are the *closed-loop state transition maps.*
- $\delta_e^s : \mathcal{X} \times \mathcal{X}^c \to \mathcal{E}_o$ for $e \in \mathcal{P}(\mathcal{E}_u \cup \mathcal{E}_d^s) - \{\emptyset\}$ are the *closed-loop output maps* (other types of events could be defined as the outputs).

- $x_0^s \in \mathcal{X}^s$ is the *initial state* of the closed-loop system.
- $\mathbf{E}_v^s$ is the set of all *valid closed-loop system event trajectories* (a subset of the event trajectories that can result given the valid plant and controller event trajectories and that the plant and controller are connected together).

The $\mathbf{E}_v^s \subset \mathbf{E}^s$, where $\mathbf{E}^s$ is the set of all infinite and finite length closed-loop input event trajectories that can be generated by the closed-loop system g^s and f_e^s. The controller has to be designed carefully so that it can eliminate the undesirable closed-loop system behavior. The $\mathbf{E}_v^s$ represents the possible behavior of the closed-loop system. Clearly, $\mathbf{E}_v$ and $\mathbf{E}_v^c$ place constraints on the allowable event sequences of the closed-loop system, that is, on $\mathbf{E}_v^s$. Furthermore, additional constraints on which event sequences are physically possible in the closed-loop system can be represented via $\mathbf{E}_v^s$. There is the need to specify the initial state of the closed-loop system to reduce the insignificant closed-loop state combinations which may unnecessarily complicate the model. If we know the initial state of the closed-loop system, the state transitions can be constrained to the possible closed-loop states only, which enables us to eliminate the impossible closed-loop state combinations. Next we define g^s and f_e^s.

Given $x_o^s = (x_o, x_o^c)$, the initial state of the closed-loop system, then if we are at state $x_k^s = (x_k, x_k^c)$ at time k, then

$$g^s(x_k^s) = [g^c(x_k^c) \cap \mathcal{E}_d^c] \cup [\delta^c(x_k^c) \cap g(x_k)] \cup [g(x_k) \cap \mathcal{E}_d],$$

where

- $g^c(x_k^c) \cap \mathcal{E}_d^c$ is the set of reference input events to the controller that are enabled at the controller state x_k^c.
- $\delta^c(x_k^c) \cap g(x_k)$ is the set of command input events to the plant that are enabled at the current controller state x_k^c and also in the current plant state x_k.
- $g(x_k) \cap \mathcal{E}_d$ is the set of disturbance input events to the plant that are enabled at the current plant state x_k.

The enabled plant input events in the closed-loop system are the plant events that are enabled by both the current plant state x_k and controller states x_k^c and the disturbance input events that are enabled at the plant state x_k. Thus, the controller can only control the enabling of events of type $\mathcal{E}_u$ only. It is assumed that the controller is appropriately defined so that it is able to transition to its next state in response to any plant output event which occurs. This is needed so that the closed-loop system dynamics are well-defined. Next we define f_e^s.

Let $e_k \subset g^s(x_k^s)$ be an enabled event in the closed-loop system at state $x_k^s = (x_k, x_k^c)$, then if

$e^c_{d_k} \in g^c(x^c_k) \cap \mathcal{E}^c_d,$
$e_{u_k} \in \delta^c(x^c_k) \cap g(x_k),$ and
$e_{d_k} \in g(x_k) \cap \mathcal{E}_d$

e_k can be given by one of the following (depending on what kind of command and disturbance input events occur simultaneously): (i) $e^1_k = \{e_{u_k}, e_{d_k}, e^c_{d_k}\}$, (ii) $e^2_k = \{e_{u_k}, e_{d_k}\}$, (iii) $e^3_k = \{e_{u_k}, e^c_{d_k}\}$, (iv) $e^4_k = \{e_{d_k}, e^c_{d_k}\}$, (v) $e^5_k = \{e_{u_k}\}$, (vi) $e^6_k = \{e_{d_k}\}$, or (vii) $e^7_k = \{e^c_{d_k}\}$. We will call e^i_k an "event of type i", where $1 \leq i \leq 7$. Let us denote $e_{p_k} = \{e_{u_k}, e_{d_k}\}$, $\hat{e}^i_k = e^i_k - \{e^c_{d_k}\}$ for $i = 1, 3, 4$ and $e^i_{c_k} = \{\delta_{\hat{e}^i_k}(x_k), e^c_{d_k}\}$, then corresponding to the type of e^i_k, $f^s_{e^i_k}(x^s_k) = x^s_{k+1}$ and $x^s_{k+1} = (x_{k+1}, x^c_{k+1})$, where

1. If e^1_k occurs, then $x_{k+1} = f_{e_{p_k}}(x_k)$ and $x^c_{k+1} = f^c_{e^1_{c_k}}(x^c_k)$.
2. If e^2_k occurs, then $x_{k+1} = f_{e_{p_k}}(x_k)$ and $x^c_{k+1} = f^c_{\{\delta_{e^2_k}(x_k)\}}(x^c_k)$.
3. If e^3_k occurs, then $x_{k+1} = f_{\{e_{u_k}\}}(x_k)$ and $x^c_{k+1} = f^c_{e^3_{c_k}}(x^c_k)$.
4. If e^4_k occurs, then $x_{k+1} = f_{\{e_{d_k}\}}(x_k)$ and $x^c_{k+1} = f^c_{e^4_{c_k}}(x^c_k)$.
5. If e^5_k occurs, then $x_{k+1} = f_{\{e_{u_k}\}}(x_k)$ and $x^c_{k+1} = f^c_{\{\delta_{e^5_k}(x_k)\}}(x^c_k)$.
6. If e^6_k occurs, then $x_{k+1} = f_{\{e_{d_k}\}}(x_k)$ and $x^c_{k+1} = f^c_{\{\delta_{e^6_k}(x_k)\}}(x^c_k)$
7. If e^7_k occurs, then $x_{k+1} = x_k$ and $x^c_{k+1} = f^c_{\{e^c_{d_k}\}}(x^c_k)$.

This completes the definition of the mathematical models used for the plant, expert controller, and closed-loop ECS.

Next we discuss the relevant timing issues for the closed-loop ECS. As is often done we assume that the occurrence of an event is instantaneous, which means that if an event occurs, it occurs in zero time and the state of the system changes in zero time. The closed-loop system event occurrences are synchronous in terms of event occurrences in the plant and controller. This means that the occurrence of any input event of the plant will instantaneously trigger the output event of the plant which is also an input to the controller. So the plant output event causes the controller state to transition to its next state at the same time as the occurrence of that plant input event. Corresponding to the new controller state there is a controller output which becomes the next enabled plant command input. However, only the events which are also enabled at the next plant state can possibly occur.

Although every occurrence of an input event of the plant *always* affects the controller state, the occurrence of an input event of the controller does not necessarily immediately affect the plant state. For instance, a reference input event of the controller, $e^c_d \in \mathcal{E}^c_d$, can occur by itself causing a controller state to transition to its next state without affecting the plant state. This can be observed in the mathematical expression of the input event of type 7 above, which affects only the controller state part of the closed-loop state $x^s = (x, x^c)$.

Note that the enabled input events of the closed-loop system can occur asynchronously as long as their event trajectories lie in $\mathbf{E}_v^s$. The $\mathbf{E}_v^s$ can be used to model the *forced plant command input* event if we add some constraints in addition to the ones corresponding to the plant and controller valid event trajectories. Some of the possible constraints are: (1) Every time the closed-loop system state changes, the next plant input event that can occur must include the plant command input enabled by the current controller state, or (2) Every time the closed-loop system state changes, the next input event that can occur must be only the plant command input event which is enabled by the current controller state.

Analysis Techniques for the ECS Suppose that a system represented with G exhibits some undesirable behavior due to, for example, its disturbance inputs e_{d_k}. The control problem involves choosing the controller C, so that, when the plant is embedded in the closed-loop system, the closed-loop system will not exhibit any undesirable behavior. To achieve its task, the controller observes the plant outputs e_{o_k} and generates command inputs e_{u_k} to the plant. The reference inputs $e_{d_k}^c$ to the controller are used to change the objectives of the controller as it operates (for example, in conventional control the reference input is often used to specify the desired value of the output). In analysis ,the focus is on testing if the plant G, controller C, and especially the closed-loop ECS S satisfy certain properties to be considered next. In our discussions below we will refer to analysis of the closed-loop properties of S but clearly the properties and analysis are also valid for G and C.

Reachability Properties: The results in [32] showed the relationship between performing chains of inference and reachability. In particular, the authors define reachability in the context of inference processes as the ability to fire a sequence of rules to derive a specific conclusion from some specific initial knowledge. In system-theoretic terms this is a standard definition for reachability that one might call a "state-to-state" property. Here we consider a slightly more general reachability property for studying inference processes in the expert controller and ECS. For $\mathcal{X}_m \subset \mathcal{X}^s$, let $\mathcal{X}(S, x_0^s, \mathcal{X}_m)$ denote the set of all finite length state trajectories that begin at x_0^s and end in $\mathcal{X}_m$.

Definition 6.1 (Reachability): *A system S is said to be "$(x_0^s, \mathcal{X}_m)$ – reachable" if there exists a sequence of events to occur that produces a state trajectory $s \in \mathcal{X}(S, x_0^s, \mathcal{X}_m)$.*

Note that $\mathcal{X}_m$ can represent the desired operating conditions of the ECS with x_0^s as its initial state. Hence, we will consider what could be called a "point-to-set" reachability problem for ECS. This general type of reachability is needed when it is possible that there are *several* valid states that can be reached from one initial state (or in the situation where it is known that at least one state in a set of states $\mathcal{X}_m$ is reachable). To automate testing of the property in Definition 6.1 we use the *A* search algorithm* [82, 69] to find the state trajectory $s \in \mathcal{X}(S, x_0^s, \mathcal{X}_m)$ when it exists.

Cyclic Properties: In the verification of the dynamic properties of the expert controller's inference loop or the ECS, the study of cyclic behavior is of paramount importance. This is due to the fact that if cycles exist, the system could get "trapped" in a *circular argument* so that there is no way it can achieve its ultimate task. This cyclic characteristic will be particularly problematic for expert control systems that operate in time-critical environments (e.g., in a process control failure diagnosis problem). Let $\mathcal{X}_c \subset \mathcal{X}^s$ denote a subset of the states such that each $x_c \in \mathcal{X}_c$ lies on a cycle that is in $\mathcal{X}_c$.

Definition 6.2 (Cyclic): *A system S is said to be "$(x_0^s, \mathcal{X}_c)$ – cyclic" if there exists a sequence of events to occur that produces a state trajectory $s \in \mathcal{X}(S, x_0^s, \mathcal{X}_c)$.*

It is a hard problem to detect the presence of cyclic behavior in the system, since one may not be able to find $\mathcal{X}_c$ without studying all system trajectories. To help automate the testing of the property in Definition 6.2 we can use a two step approach. First we specify a set $\mathcal{X}_c$ (which can sometimes be found with a search algorithm), then we use a search algorithm to find the inference path that starts at x_0^s and ends in $\mathcal{X}_c$ (if one exists) [71]. This approach is used in [49] to study cyclic properties of an expert system. In our applications in Sections 5 and 6 we will actually verify that the ECS does not contain undesirable cycles by verifying certain stability properties.

Stability Properties: In terms of characterizing human cognitive functions, Lyapunov stability for the expert controller can be viewed as a mathematical characterization of an expert controller's ability to concentrate (i.e., to focus, to pay attention) on the control problem. From an engineering, rather than psychological standpoint, stability of a control system is of fundamental importance due to the fact that guarantees of stability often ensure that the system variables will stay in *safe* operating regions and that other performance objectives can be met. In this case study we will show how some of the stability properties defined in Chapter 3 can be studied in a Lyapunov framework.

An important advantage of the Lyapunov approach in the study of stability properties is that it is often possible to intuitively define an appropriate Lyapunov function (years of use have shown this). However, specifying the Lyapunov function is sometimes a difficult task. Motivated by the difficulties in specifying a Lyapunov function, one can use search algorithms to study stability properties. The study of asymptotic stability in the large or of regions of asymptotic stability using search methods is a three-step process. First, we have to find the invariant set $\mathcal{X}_m$ and then determine the region of asymptotic

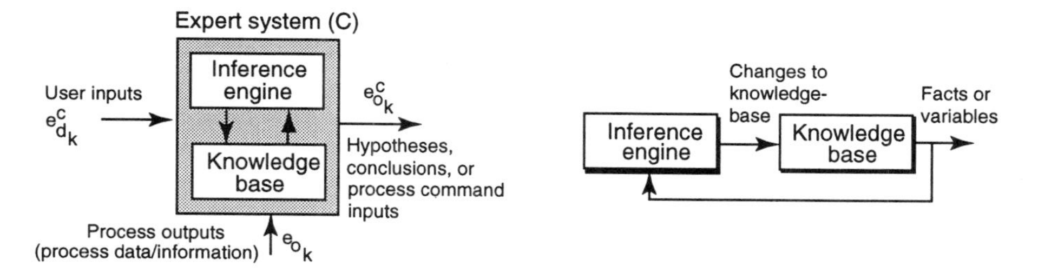

Fig. 6.2 An expert system viewed as a control system

stability $\mathcal{X}_v$. The third step is to show that all paths which originate from any state in $\mathcal{X}_v$ will end up in $\mathcal{X}_m$. We can modify the A* algorithm [82] to expand all successor states of a given initial state $x_0^s \in \mathcal{X}_v$ until the "open" set is empty. If all paths originating from any $x_0^s \in \mathcal{X}_v$ converge to the set $\mathcal{X}_m$, then $\mathcal{X}_m$ is asymptotically stable w.r.t. $\mathbf{E}_v^s$ for the region $\mathcal{X}_v$ (one must be careful with cycles, dead-end states, and the imposition of the constraints specified by $\mathbf{E}_v^s$). We briefly mention our use of search algorithms for the analysis of stability properties for the applications later in this chapter.

6.2.3 Expert System Verification

In this Section we discuss how to analyze the properties of the isolated expert controller (i.e., the expert system that implements the controller *without* the plant). In Figure 6.2 we show a general expert system that interfaces to "user inputs" and "process outputs" and makes decisions about what "process inputs" to generate. The second part of Figure 6.2 shows the central part of the expert system redrawn to highlight the fact that it is a control system; in this case: (1) the "plant" is the knowledge-base, (2) the "controller" is the inference engine, (3) the "command inputs" are the changes that the inference engine makes to the knowledge-base, and (4) the "outputs" of the closed-loop system are the facts/variables in working memory (that the inference engine uses in its decision-making process). One can define the "reference inputs" in several different ways (e.g., they could represent "goal inputs" to which the inference engine can react). In [49] the authors show how to specify a plant model for the knowledge-base and a controller model for the inference engine (i.e., the inference loop shown on the right of Figure 6.2), find the closed-loop system model, and analyze qualitative properties of a particular expert system.

In [50] the authors show that it is actually possible, independent of the plant, to characterize several general properties of the expert controller and closed-loop ECS. For instance, for the isolated expert controller any invariant set $\mathcal{X}_m \subset \mathcal{X}^c$ is stable in the sense of Lyapunov w.r.t. $\mathbf{E}_v^c$ and asymptotically stable w.r.t. $\mathbf{E}_v^c$ since these are *local* properties. It is not necessarily the case that these properties hold trivially for the full ECS, due to the dynamics of the plant. Note that it may be useful to analyze asymptotic stability in the large for the isolated expert controller to show that in the limit, the state of the expert controller will reach some set of states. Moreover, it is proven in [76] that the expert controller C is, in general, unbounded w.r.t. $\mathbf{E}_v^c$ and any fixed bounded set $\mathcal{X}_b \subset \mathcal{X}^c$. This shows that precautions must be taken to ensure that the expert controller does not become unbounded, as this will cause problems with overflow in the implementation of the expert controller. To avoid this problem one can simply remove the "recency" conflict-resolution strategy. Alternatively, one must be careful to make sure that the rule-base is designed so that rules cannot enter the conflict set and stay there without eventually being fired or having their premise data changed.

6.2.4 Expert Control of a Surge Tank

The surge tank has two fill valves labeled as "A" and "B" as illustrated in Figure 6.3. An empty valve C is located at the bottom of the tank. The tank contains liquid whose level is denoted by a nonnegative integer. It is assumed that we have complete control of the opening of both valves A and B. When it is opened, valve A automatically closes itself after pouring enough liquid so that the liquid level in the tank increases by one level. Similarly, when valve B is opened, it automatically shuts itself down after pouring enough liquid so that the liquid level in the tank increases by three levels. We require that only one fill valve can be opened at once. The opening of valve C is random and unpredictable; however, it is assumed that for all times there always exists another time that valve C will open (i.e., C will be persistently opened). Once valve C opens, the liquid level in the tank decreases by two (if the liquid level was greater than or equal to 2) or until empty (if the liquid level was less than 2), and then it automatically closes itself. It is assumed that there is a sensor which can measure the height of the liquid in the tank and provide its measured samples in an asynchronous fashion. The control objective for the expert controller in this example is to regulate the liquid level in the tank so that it lies in the range between 2 (the minimum safety level) and 7 for any given initial liquid level. Ideally, however, to reduce the magnitude of the variations in the liquid level we would like the liquid level to eventually lie only in the range between 2 and 5.

The plant characteristics can be modeled via G in Equation (6.1) where: (1) the set of plant states is $\mathcal{X} = \{0, 1, 2, \ldots\}$ where $x \in \mathcal{X}$ denotes the liquid level in the tank, (2) the set of plant command input events is $\mathcal{E}_u = \{e_a, e_b, e_{nil}\}$ where e_a denotes valve A is opened, e_b denotes valve B is opened, and e_{nil} denotes valves A and B are kept closed, (3) the set of plant disturbance input events is $\mathcal{E}_d = \{e_c\}$ where e_c denotes valve C is opened, and (4) we have $\mathcal{E}_o = \mathcal{X}$, $\delta_e(x) = x$, and $g(x) = \{e_a, e_b, e_{nil}, e_c\}$ for all $x \in \mathcal{X}$. Suppose that $e_i \subset g(x)$, $1 \leq i \leq 7$, denote various events that can occur at state x. Then, $e_1 = \{e_a\}$, $e_2 = \{e_b\}$, $e_3 = \{e_c, e_{nil}\}$, $e_4 = \{e_c, e_a\}$, $e_5 = \{e_c, e_b\}$, $e_6 = \{e_c\}$,

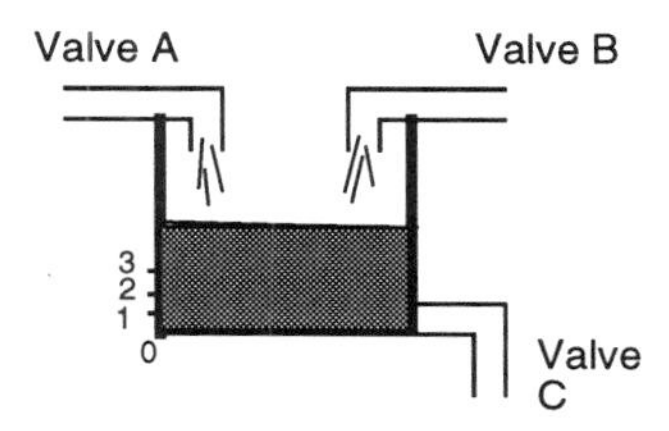

Fig. 6.3 Surge tank.

$e_7 = \{e_{nil}\}$, so that $f_{e_1}(x) = x + 1$, $f_{e_2}(x) = x + 3$, $f_{e_3}(x) = f_{e_6}(x)$, $f_{e_3}(x) = x - 2$ if $x \geq 2$, $f_{e_3}(x) = x - 1$ if $x = 1$, $f_{e_3}(x) = 0$ if $x = 0$, $f_{e_4}(x) = x - 1$ if $x \geq 1$, $f_{e_4}(x) = 0$ if $x = 0$, $f_{e_5}(x) = x+1$, and $f_{e_7}(x) = x$. Finally, let $\mathbf{E}_v \subset \mathbf{E}$ be all event trajectories composed of events e_i such that $i \neq 6, 7$; hence, we assume that the command input events will always accompany a disturbance input event if it occurs and we assume that e_{nil} cannot occur by itself. Note that time progresses in an asynchronous fashion.

Modeling and Design of the Expert Controller for the Surge Tank One strategy that a human expert can use to meet the control objectives is:

1. Drive the liquid level in the tank from its initial level to 5 by enabling e_b when the liquid level is less than or equal to 4 (to ensure the level will increase regardless of the occurrence of e_5) or e_{nil} when the liquid level is higher than 4 (to decrease the liquid level).

2. When the liquid level is at 5, enable e_{nil} so that the level will not exceed 5. The level will eventually drop to 3, due to an occurrence of e_3 or e_6.

3. When the level is at 4, we do not need to increase the level since it is already within the ideal range. Thus, enable e_{nil}. The level will eventually drop to 2, due to an occurrence of e_3 or e_6.

4. When the liquid level is at 3, enable e_a so that the level may increase without exceeding level 5. The level will eventually increase to 4, due to an occurrence of e_1 or drop to 2, due to an occurrence of e_4.

5. When the level is at 2, enable e_b to ensure that the liquid level will increase regardless of the occurrence of e_2 or e_5. The level will eventually increase to 5, due to an occurrence of e_2 or to 3, due to an occurrence of e_5.

Use of such a strategy inherently depends on the plant model so we first use the model-based approach to expert controller design. The controller can be modeled using C in Equation (6.2), where:

- The set of knowledge-base states is $\mathcal{X}^b$, where

$$\boldsymbol{x}^b = [x_1^b, x_2^b, x_3^b, x_4^b, x_5^b, x_6^b]^\top \in \mathcal{X}^b$$

and x_1^b, x_2^b are variables and $x_3^b, x_4^b, x_5^b, x_6^b$ are facts, and x_1^b represents the current liquid level in the tank, $x_2^b = 0, 1, 2$, indicating that e_{nil}, e_a, e_b is enabled respectively, $x_3^b = T(a_1) = 0$, indicating that the liquid level has never reached 5, $x_3^b = T(a_1) = 1$ indicating that the liquid level has reached 5; $x_4^b = T(a_2) = 0$, indicating that rules 1 and 2 are allowed to fire (to be defined below), $x_4^b = T(a_2) = 1$, indicating that rules 1 and 2 are not allowed to fire; $x_5^b = T(a_3) = 0$, indicating that rules 3 and 4 are allowed to fire, $x_5^b = T(a_3) = 1$, indicating that rules 3 and 4 are not allowed to fire; $x_6^b = T(a_4) = 0$, indicating that rule 5 is allowed to fire, $x_6^b = T(a_4) = 1$, indicating that rule 5 is not allowed to fire;

- The set of inference engine states is $\mathcal{X}^i$, where $\boldsymbol{x}^i = [\boldsymbol{p}^\top \ \boldsymbol{s}^\top]^\top \in \mathcal{X}^i$ and $\boldsymbol{p}$ and $\boldsymbol{s}$ are 7×1 vectors.
- The set of expert controller command inputs is $\mathcal{E}_u^\ell = \mathcal{E}_o = \mathcal{X}$. Also, $\mathcal{E}_d^c = \emptyset$, $\mathcal{E}_o^c = \mathcal{E}_u$, and $\mathbf{E}_v^c = \mathbf{E}^c$.
- The $g^c(x^c)$ and $f_e^c(x^c)$ are defined via rules $r_i \in \mathcal{R}$, $i = 1, 2, \ldots, 7$, which are:
  - r_1: IF $(x_1^b \leq 4 \wedge T(a_1) \wedge \neg T(a_2) \wedge e_u^\ell \leq 2)$ THEN $x_1^b := e_u^\ell \wedge x_4^b := 1 \wedge x_5^b := 0 \wedge x_6^b := 0 \wedge x_2^b := 2$
  - r_2: IF $(x_1^b \leq 4 \wedge T(a_1) \wedge \neg T(a_2) \wedge e_u^\ell > 2)$ THEN $x_1^b := e_u^\ell \wedge x_4^b := 1 \wedge x_5^b := 0 \wedge x_6^b := 0 \wedge x_2^b := 0$
  - r_3: IF $(x_1^b \leq 2 \wedge T(a_1) \wedge \neg T(a_3) \wedge e_u^\ell \leq 4)$ THEN $x_1^b := e_u^\ell \wedge x_4^b := 0 \wedge x_5^b := 1 \wedge x_6^b := 0 \wedge x_2^b := 1$
  - r_4: IF $(x_1^b \leq 2 \wedge T(a_1) \wedge \neg T(a_3) \wedge e_u^\ell > 4)$ THEN $x_1^b := e_u^\ell \wedge x_4^b := 0 \wedge x_5^b := 1 \wedge x_6^b := 0 \wedge x_2^b := 0$
  - r_5: IF $(x_1^b = 5 \wedge \neg T(a_4) \wedge e_u^\ell \geq 0)$ THEN $x_1^b := e_u^\ell \wedge x_3^b := 1 \wedge x_6^b := 1 \wedge x_2^b := 1$
  - r_6: IF $(x_1^b \geq 0 \wedge e_u^\ell \leq 4)$ THEN $x_1^b := e_u^\ell \wedge x_2^b := 2$
  - r_7: IF $(x_1^b \geq 0 \wedge e_u^\ell > 4)$ THEN $x_1^b := e_u^\ell \wedge x_2^b := 0$
  - (Note that some extra premise functions are added merely to raise the distinctiveness of a rule.)
- The output event function is defined as $\delta^c(x^c) = x_2^b$

The $T(a_i)$ for $i = 1, 2, 3, 4$ are facts containing "flags," which emulate human expert reasoning sequences in controlling the surge tank. All the conflict resolution strategies are utilized in this example in the order they were presented earlier. Since firing any rule causes x_1^b to be updated and the antecedents of all the rules test x_1^b, the matrix $\boldsymbol{A}$ is a 7×7 matrix of ones. Thus, in this problem, refraction and recency do not prune rules from the conflict set. This actually ensures that the recency conflict-resolution strategy cannot cause the expert controller to become unbounded. The distinctiveness level is automatically specified once the rules are loaded into the knowledge-base (by the number of terms in the left hand sides of the rules) and the priority levels of the rules are all defined to have the same value. Hence, it happens that this example actually utilizes the conflict-resolution strategies that prune rules from the conflict set based only on distinctiveness (and arbitrary).

The initial closed-loop state for the model S of the full ECS (which can be specified as explained in Section 3) is $x_0^s \in \mathcal{X}_0^s$, where $\mathcal{X}_0^s \subset \mathcal{X}^s$ and

$$\begin{aligned}\mathcal{X}_0^s &= \{x^s \in \mathcal{X}^s : x_1^b = x, x_2^b = 2 \text{ if } x \leq 4, x_2^b = 0 \\ &\quad \text{if } x > 4, x_i^b = 0, i = 3, 4, 5, 6, \mathbf{p} = \mathbf{s} = 0\}.\end{aligned}$$

The set of initial states $\mathcal{X}_0^s$ indicates that the initial liquid level in the tank can be arbitrary (but the expert controller must know what it is); however, the input event to the plant e_b must be enabled initially when the level is less than or equal to 4, and e_{nil} must be enabled if the level is greater than 4. The valid closed-loop ECS state trajectories $\mathbf{E}_v^s$ are defined to be the state trajectories which can be generated by g, f_e, g^c and f_e^c, and which satisfy the $\mathbf{E}_v$ and $\mathbf{E}_v^c$ constraints.

Analysis It is obvious that there exist cycles in the open-loop plant. For example, if the initial liquid level in the tank is 10, then corresponding to the occurrence of the event trajectory "$e_a e_c e_a$," the state trajectory has a liquid level sequence of "10, 11, 9, 10." Maintaining the liquid level around 5 may not be possible in the open-loop plant, since the state trajectories of the plant may exhibit cycles outside the desired set of states. Furthermore, the liquid level can be unbounded, corresponding to certain input event trajectories, e.g., infinite number of occurrences of e_a, e_b, or their combinations in sequence. In order to eliminate these undesirable properties and meet the closed-loop control specifications, the expert controller above is employed. The analysis of the closed-loop ECS for the surge tank is illustrated next.

When closed-loop expert control is used for this surge tank, the set

$$\mathcal{X}_{st} = \{x^s \in \mathcal{X}^s : x \in \{2, 3, 4, 5, 6, 7\}\} \tag{6.9}$$

can be shown to be invariant by simply showing that if $x_k^s \in \mathcal{X}_{st}$ it will always stay in $\mathcal{X}_{st}$. A search algorithm (the $A*$ algorithm) was used to study reachability properties of the ECS for the surge tank. Using this algorithm we show that there exists at least one path from any given initial liquid level in the tank (within certain initial liquid level bounds) which leads to a state indicating that the liquid level is 5. The results of our reachability analysis are stated in the following result.

Theorem 6.1 (Reachability for the Surge Tank): *The ECS for the surge tank described above is $(x_0^s, \mathcal{X}_{st})$-reachable for all $x_0^s \in \mathcal{X}_0^s$, since there exists a sequence of events to occur that produces a state trajectory $s \in \mathcal{X}(S, x_0^s, \mathcal{X}_{st})$ for any $x_0^s \in \mathcal{X}_0^s$.*

This reachability result shows that the expert controller can make appropriate "chains of inference" to reason about how to control the plant. We have, however, not yet shown that the control objectives related to liquid level regulation are achieved (we have shown that the trajectories exist; we are not guaranteed that the expert controller will follow them). This is addressed next.

Let the distance between two states x^s and $x^{s'}$ be defined as

$$\rho(x^s, x^{s'}) = \sqrt{(x - x')^2 + \sum_{i=1}^{6}(x_i^b - x_i^{b'})^2 + \sum_{j=1}^{14}(x_j^i - x_j^{i'})^2}, \tag{6.10}$$

where $x^s = (x, \boldsymbol{x}^b, \boldsymbol{x}^i)$ and $x^{s'} = (x', \boldsymbol{x}^{b'}, \boldsymbol{x}^{i'})$.

Theorem 6.2 (Region of Asymptotic Stability for the Surge Tank): *The invariant set $\mathcal{X}_{st}$ for the surge tank above has a region of asymptotic stability w.r.t. $\mathbf{E}_v^s$ of $\mathcal{X}_0^s$.*

Proof: From Equation (6.10) and from the definition of $\mathcal{X}_{st}$ in Equation (6.9), $\rho(x^s, \mathcal{X}_{st})$ can be simplified to

$$\rho(x^s, \mathcal{X}_{st}) = \min\{|x - \bar{x}| : \bar{x} \in \{2, 3, 4, 5, 6, 7\}\}. \tag{6.11}$$

Define the Lyapunov function $V(x^s) = \rho(x^s, \mathcal{X}_{st})$, so that choosing $c_1 = c_2 = 1$ we get $c_1\rho(x^s, \mathcal{X}_{st}) \leq V(x^s) \leq c_2\rho(x^s, \mathcal{X}_{st})$, which results in the satisfaction of condition (i) of Theorem 3.1. We want to show that if the initial state $x_0^s \in \mathcal{X}_0^s$, $V(x_k^s)$ is a nonincreasing function in k and $V(x_k^s) \to 0$ as $k \to \infty$ for the rules that the expert controller can fire. From the rule-base, it is obvious that for any initial state $x_0^s \in \mathcal{X}_0^s$, the only rules which can fire are rules 5, 6, and 7. There are three possible cases:

Case 1: The state x^s has $x < 2$: For this case, rule 6 is the only rule which is enabled and can be fired. The firing of this rule enables e_b. Thus, the only input events to the plant which can occur are $\{e_b, e_c\}$ and $\{e_b\}$. The liquid level in the tank will definitely increase regardless of the occurrence of $\{e_b, e_c\}$ or $\{e_b\}$ causing $V(x_k^s)$ to decrease, with time k. Once the liquid reaches a level that is greater than or equal to 2, consider Case 3.

Case 2: The state x^s has $x > 7$: For this case, rule 7 is the only rule which is enabled and can be fired. The firing of this rule enables e_{nil}. Thus, the only input event to the plant which can occur is $\{e_c, e_{nil}\}$. The liquid level in the tank will definitely decrease causing $V(x_k^s)$ to decrease with time k. Once $x \leq 7$, consider Case 3.

Case 3: The state x^s has $2 \leq x \leq 7$: Once the state x^s has $2 \leq x \leq 7$, the invariant set $\mathcal{X}_{st}$ is reached. Thus, $V(x_k^s) = 0$.

This proves that $V(x_k^s)$ is a nonincreasing function in k and $V(x_k^s) \to 0$ as $k \to \infty$. ■

The ECS for the surge tank is not asymptotically stable in the large, because the initial state of the knowledge-base and inference engine must be appropriate. In other words, the expert controller must have the correct initial knowledge about the liquid level in the tank and the knowledge-base and inference engine states must be initialized properly, depending on the initial liquid level in the tank. Also, note that it is easy to see that there exists a $k' \geq 0$, such that $x_{k''} \in \{2,3,4,5\}$ for all $k'' \geq k'$; hence this expert controller meets the "ideal" design objectives listed above. Finally, we note that we have also used search algorithms to show that the ECS does not exhibit cyclic behavior outside the invariant set X_{st} (of course we also know this from the stability analysis above). Search algorithms can also be used to verify asymptotic stability for this application.

Discussion: Design Issues Clearly, different closed-loop system behavior will result from different rule-bases and inference mechanisms. For instance, another expert controller can be designed so that the resulting ECS exhibits different behavior within the same invariant set $\mathcal{X}_{st}$ (but the invariant set $\mathcal{X}_{st}$ still possesses the same stability properties). This controller has a simpler knowledge-base structure which utilizes only 3 rules. In this case, the plant state is not included in the knowledge-base. The controller will act solely based on the liquid level in the surge tank. No conflict resolution strategies are needed. The controller has a one dimensional state $x^c = x^b$ corresponding to the enabled command input to the plant. The output map is defined as $\delta^c(x^c) = x^b$. The rules are as follows:

- r_1 : IF $e_u^\ell \leq 2$ THEN $x^b := 2$
- r_2 : IF $e_u^\ell \geq 3 \wedge e_u^\ell \leq 6$ THEN $x^b := 1$
- r_3 : IF $e_u^\ell > 6$ THEN $x^b := 0$

Due to the simplicity of the rule-base structure, it is obvious that the same stability properties for the invariant set $\mathcal{X}_{st}$ can also be achieved using this controller. The trade-off between this controller and the 7-rule controller is that the state trajectories in the invariant set are *less restrictive* for the 3-rule controller. The 7-rule controller drives the initial state of the plant to a state where $x = 5$, and, once it is there, it follows only the state trajectories $5,3,2,5,\ldots$; $5,3,2,3,5,\ldots$; $5,3,4,2,5,\ldots$; and $5,3,4,2,3,5,\ldots$. For the 3-rule controller there are many possible variations in the state trajectories, depending on the first state reached in the invariant set $\mathcal{X}_{st}$. In fact, the 3-rule controller does not satisfy the *ideal* control specifications (stated above), since it cannot eventually keep the level between 2 and 5. The plant state is included in the 7-rule controller design and because the information about the current and the next liquid level in the tank is used, an appropriate command input can be enabled to produce the state trajectories mentioned above. The 7-rule controller is analogous to a *dynamic* controller in conventional control systems, since the 7-rule controller uses memory when making decisions regarding the enabling of command input events to the plant. On the other hand, the 3-rule controller is analogous to *static* controller in conventional control systems,

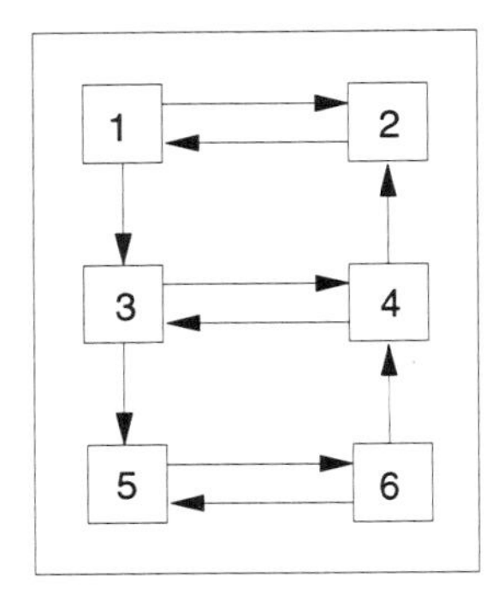

Fig. 6.4 Six-machine FMS.

since it is simply a nonlinear mapping between the controller's input and output. Overall, this example serves to illustrate the trade-off in using a dynamic controller with more rules, as opposed to a simple static controller with fewer rules: using more rules we are able to achieve more demanding specifications.

6.2.5 Expert Control of a Flexible Manufacturing System

The flexible manufacturing system (FMS) that we consider here is composed of a set of identical machines connected by a robotic transportation system. The FMS can be represented by a directed graph (M, T) where $M = \{1, 2, \ldots, N\}$ represents a set of machines numbered by $i \in M$, and $T \subset M \times M$ is the set of transportation tracks between machines. We assume that (M, T) is *strongly connected*, that is, that for any $i \in M$ there exists a path from i to every other $j \in M$ (also, if $(i, j) \in T$, $i \neq j$). This ensures that no machine is isolated from any other machine in the FMS. Each machine has a queue that holds parts that can be processed by any machine in the system. Let the number of parts in the queue of machine $i \in M$ be denoted by $x_i \geq 0$. There are robotic transporters that travel on the tracks represented by $(i, j) \in T$ and move parts between the queues of various machines. The robots can transfer parts from any $i \in M$ to another $j \in M$ if there exists a track $(i, j) \in T$. Let m_{ij} denote the command input to the robot on $(i, j) \in T$ to "move parts from machine i to machine j". In this section we focus on expert control of the FMS with six machines, and track topology shown in Figure 6.4. The arrows indicate the directions parts can be transferred. For example, machine 1 can pass its parts to machines 2 and 3 via the robots (due to an m_{12} or m_{13} command, respectively), and machine 2 can pass its parts only to machine 1 (due to an m_{21} command).

The problem we focus on is how to move the robotic transporters on the tracks $(i, j) \in T$, so that they can redistribute the parts in the queues of the machines so that all machines are equally loaded (so that under-utilization of processing resources is not encountered). It is assumed that the robot knows the graph (M, T), and can sense the number of parts in each machine $i \in M$, and that the initial total number of parts satisfy

$$\sum_{i=1}^{6} \frac{x_i}{6} = R \in \mathbb{N}, \tag{6.12}$$

(where $\mathbb{N}$ denotes the set of nonnegative integers) so that we seek to balance to R parts for each machine $i \in M$ (lack of satisfaction of this condition makes it possible to achieve only an imperfect type of balancing). Furthermore, it is assumed that no new parts arrive from outside the FMS and that no parts are processed by the machines while the redistribution takes place (if parts arrive and depart slower than the redistribution takes place, our expert controller can be modified so that it will persistently seek to balance the load). The number of parts that can be transported from machine i to machine j in one travel is defined to be $\left|\frac{x_i - x_j}{2}\right|$ if $|x_i - x_j|$ is divisible by 2, and the number of parts moved is less than or equal to x_i, or $\left|\frac{x_i - x_j}{2}\right| + \frac{1}{2}$ if $|x_i - x_j|$ is not divisible by 2, and the number of parts moved is less than or equal to x_i, or x_i, otherwise (note that this is *without* control). The FMS described above can be modeled by G, where:

- The plant states are $\boldsymbol{x} = [x_1, x_2, \ldots, x_6]^\top \in \mathcal{X}$ with x_i the number of parts in the queue of $i \in M$
- The set of plant command input events is

  $$\mathcal{E}_u = \{m_{00}, m_{12}, m_{13}, m_{21}, m_{35}, m_{34}, m_{42}, m_{43}, m_{56}, m_{65}, m_{64}\},$$

  where m_{00} denotes the event "no part is moved."
- We have $\mathcal{E}_d = \emptyset$ and $\mathcal{E}_o = \mathcal{X}$.
- $g(\boldsymbol{x})$ and $f_e(\boldsymbol{x})$ are defined below, where $\boldsymbol{x} = [x_1\ x_2\ \ldots\ x_6]^\top$ and $\boldsymbol{x}' = [x_1'\ x_2'\ \ldots\ x_6']^\top$ are the current and next plant states, respectively. Note that, after the occurrence of event m_{ij},

  $$x_i' = \begin{cases} x_i - \left|\frac{x_i - x_j}{2}\right| & \text{if } \left|\frac{x_i - x_j}{2}\right| = r \leq x_i \text{ and } r \in \mathbb{N} \\ x_i - \left(\left|\frac{x_i - x_j}{2}\right| + \frac{1}{2}\right) & \text{if } \left|\frac{x_i - x_j}{2}\right| + \frac{1}{2} = r \leq x_i \text{ and } r \in \mathbb{N} \\ 0 & \text{otherwise,} \end{cases} \tag{6.13}$$

  and

  $$x_j' = \begin{cases} x_j + \left|\frac{x_i - x_j}{2}\right| & \text{if } \left|\frac{x_i - x_j}{2}\right| = r \leq x_i \text{ and } r \in \mathbb{N} \\ x_j + \left|\frac{x_i - x_j}{2}\right| + \frac{1}{2} & \text{if } \left|\frac{x_i - x_j}{2}\right| + \frac{1}{2} = r \leq x_i \text{ and } r \in \mathbb{N} \\ x_j + x_i & \text{otherwise.} \end{cases} \tag{6.14}$$

The $x_p' = x_p$ for $p \neq i, j$.

- The output event function is defined by $\delta_e(\boldsymbol{x}) = \boldsymbol{x}$, for all e, and $\mathbf{E}_v = \mathbf{E}$.

Modeling and Design of the Expert Controller for the FMS Some knowledge that we have about how to balance the part distribution is:

1. Move parts from the queue of the machine which has the maximum number of parts to the queue of another machine which has fewer parts.

2. If machine 1 has the maximum number of parts in its queue and machines 2 and 3 have fewer parts than machine 1, transfer parts from machine 1 to machine 3 instead of to machine 2. This is to avoid transferring parts back from machine 2 to machine 1, since machine 2 can only pass its parts to machine 1.

3. If machine 6 has the maximum number of parts in its queue and machines 4 and 5 have fewer parts than machine 6, transfer parts from machine 6 to machine 4 instead of to machine 5. This is to avoid transferring parts back from machine 5 to machine 6, since machine 5 can only pass its parts to machine 6.

4. Transferring parts from machine i to j cannot be *directly* followed by transferring parts from machine j to i, and vice versa. This is to avoid passing parts back and forth between two machines.

5. After the robot transfers parts from the queue of machine 3 to the one of machine 4, it cannot repeat this transfer until it transfers parts from the queue of machine 3 to the one of machine 5. This is to avoid transferring parts only among machines 1, 2, 3, and 4 (i.e., in a cyclic fashion).

6. After the robot transfers parts from the queue of machine 4 to the one of machine 3, it cannot repeat this transfer until it transfers parts from the queue of machine 4 to the one of machine 2. This is to avoid transferring parts only among machines 3, 4, 5, and 6.

7. Transferring parts from machine i to j cannot be *directly* followed by transferring parts from machine i to j again. This is to prevent the robot from creating a less balanced distribution.

For knowledge-base design we use rules to characterize such strategies on how to balance the load. The conflict resolution strategy used here prunes rules from the conflict set, based on distinctiveness, priority (and arbitrary). Refraction and recency prunings are not utilized. The controller can be modeled using C in Equation (6.2), where:

- The set of knowledge-base states is $\mathcal{X}^b$, where $\mathbf{x}^b = [x_1^b, x_2^b, \ldots, x_{11}^b]^\top \in \mathcal{X}^b$ and x_1^b is a variable and $x_i^b, i = 2, 3, \ldots, 11$ represents the truth values of facts where $x_1^b = 0$, indicating that m_{00} is enabled, 1, indicating that

m_{12} is enabled, 2, indicating that m_{13} is enabled, 3, indicating that m_{21} is enabled, 4, indicating that m_{35} is enabled, 5, indicating that m_{34} is enabled, 6, indicating that m_{42} is enabled, 7, indicating that m_{43} is enabled, 8, indicating that m_{56} is enabled, 9, indicating that m_{65} is enabled, or 10, indicating that m_{64} is enabled; $x_i^b = T(a_{i-1}) = 0$, indicating that rule $i-1$ is allowed to fire for $i = 2, 3, \ldots, 11$, $= 1$, indicating that rule $i-1$ is not allowed to fire for $i = 2, 3, \ldots, 11$.

- There is no state associated with the inference engine so that $\mathcal{X}^c = \mathcal{X}^b$.
- We have $\mathcal{E}_u^\ell = \mathcal{E}_o = \mathcal{X}$, $\mathcal{E}_d^c = \emptyset$, and $\mathcal{E}_o^c = \mathcal{E}_u$.
- The $g^c(x^c)$ and $f_e^c(x^c)$ are defined via the rules $r_i \in \mathcal{R}$, $i = 1, 2, \ldots, 11$ given below where: P_i tests "$x_i \geq x_j \ \forall j \in M$, for $i = 1, 2, 3, 4, 5, 6$; P_7 tests "$x_1 \neq x_2$", P_8 tests "$x_1 \neq x_3$", P_9 tests "$x_3 \neq x_5$", P_{10} tests "$x_3 \neq x_4$", P_{11} tests "$x_4 \neq x_2$", P_{12} tests "$x_5 \neq x_6$", P_{13} tests "$x_6 \neq x_4$", P_{14} tests "$x_1 \geq 0$", P_{15} tests "$x_1 = x_2 \wedge x_1 = x_3 \wedge x_1 = x_4 \wedge x_1 = x_5 \wedge x_1 = x_6$",
  - C_1 means "$x_1^b := 1 \wedge x_3^b := 0 \wedge x_5^b := 0 \wedge x_7^b := 0 \wedge x_9^b := 0 \wedge x_{10}^b := 0 \wedge x_{11}^b := 0 \wedge x_2^b := 1$",
  - C_2 means "$x_1^b := 2 \wedge x_2^b := 0 \wedge x_4^b := 0 \wedge x_5^b := 0 \wedge x_7^b := 0 \wedge x_9^b := 0 \wedge x_{10}^b := 0 \wedge x_{11}^b := 0 \wedge x_3^b := 1$",
  - C_3 means "$x_1^b := 3 \wedge x_3^b := 0 \wedge x_5^b := 0 \wedge x_7^b := 0 \wedge x_9^b := 0 \wedge x_{10}^b := 0 \wedge x_{11}^b := 0 \wedge x_4^b := 1$",
  - C_4 means "$x_1^b := 4 \wedge x_2^b := 0 \wedge x_3^b := 0 \wedge x_4^b := 0 \wedge x_6^b := 0 \wedge x_7^b := 0 \wedge x_9^b := 0 \wedge x_{10}^b := 0 \wedge x_{11}^b := 0 \wedge x_5^b := 1$",
  - C_5 means "$x_1^b := 5 \wedge x_2^b := 0 \wedge x_3^b := 0 \wedge x_4^b := 0 \wedge x_5^b := 0 \wedge x_7^b := 0 \wedge x_9^b := 0 \wedge x_{10}^b := 0 \wedge x_{11}^b := 0 \wedge x_6^b := 1$",
  - C_6 means "$x_1^b := 6 \wedge x_2^b := 0 \wedge x_3^b := 0 \wedge x_4^b := 0 \wedge x_5^b := 0 \wedge x_8^b := 0 \wedge x_9^b := 0 \wedge x_{10}^b := 0 \wedge x_{11}^b := 0 \wedge x_7^b := 1$",
  - C_7 means "$x_1^b := 7 \wedge x_2^b := 0 \wedge x_3^b := 0 \wedge x_4^b := 0 \wedge x_5^b := 0 \wedge x_7^b := 0 \wedge x_9^b := 0 \wedge x_{10}^b := 0 \wedge x_{11}^b := 0 \wedge x_8^b := 1$",
  - C_8 means "$x_1^b := 8 \wedge x_2^b := 0 \wedge x_3^b := 0 \wedge x_4^b := 0 \wedge x_5^b := 0 \wedge x_7^b := 0 \wedge x_{11}^b := 0 \wedge x_9^b := 1$",
  - C_9 means "$x_1^b := 9 \wedge x_2^b := 0 \wedge x_3^b := 0 \wedge x_4^b := 0 \wedge x_5^b := 0 \wedge x_7^b := 0 \wedge x_{11}^b := 0 \wedge x_{10}^b := 1$",
  - C_{10} means "$x_1^b := 10 \wedge x_2^b := 0 \wedge x_3^b := 0 \wedge x_4^b := 0 \wedge x_5^b := 0 \wedge x_7^b := 0 \wedge x_9^b := 0 \wedge x_{10}^b := 0 \wedge x_{11}^b := 1$",
  - C_{11} means "$x_1^b := 0$".
- The rules $r_i \in \mathcal{R}$ are given by
  - r_1 : IF $P_1 \wedge \neg T(a_1) \wedge \neg T(a_3) \wedge P_7$ THEN C_1

- r_2 : IF $P_1 \wedge \neg T(a_2) \wedge P_8 \wedge P_{14}$ THEN C_2
- r_3 : IF $P_2 \wedge \neg T(a_3) \wedge \neg T(a_1) \wedge P_7$ THEN C_3
- r_4 : IF $P_3 \wedge \neg T(a_4) \wedge P_9 \wedge P_{14}$ THEN C_4
- r_5 : IF $P_3 \wedge \neg T(a_5) \wedge P_{10} \wedge P_{14}$ THEN C_5
- r_6 : IF $P_4 \wedge \neg T(a_6) \wedge P_{11} \wedge P_{14}$ THEN C_6
- r_7 : IF $P_4 \wedge \neg T(a_7) \wedge P_{10} \wedge P_{14}$ THEN C_7
- r_8 : IF $P_5 \wedge \neg T(a_8) \wedge \neg T(a_9) \wedge P_{12}$ THEN C_8
- r_9 : IF $P_6 \wedge \neg T(a_9) \wedge \neg T(a_8) \wedge P_{12}$ THEN C_9
- r_{10} : IF $P_6 \wedge \neg T(a_{10}) \wedge P_{13} \wedge P_{14}$ THEN C_{10}
- r_{11} : IF P_{15} THEN C_{11}

- The output event function is defined by $\delta^c(x^c) = x_1^b$ and $\mathbf{E}_v^c = \mathbf{E}^c$.

The $T(a_i)$ for $i = 1, 2, \ldots, 10$ are facts containing "flags" which emulate human expert reasoning sequences in balancing part distribution in a FMS. These flags are set to 0 or 1, based on rule firing sequences and in fact implement a special type of "refraction" conflict-resolution strategy (hence the inference strategies in the inference engine can be disabled and effectively replaced by appropriate ones implemented in the knowledge-base). For example, due to the presence of $T(a_5)$ in the antecedents of rule 5, rule 5 cannot be fired again after the firing of rule 5 until the firing of rule 4. The conflict resolution strategies used here are distinctiveness, priority (and arbitrary). The distinctiveness level is automatically defined as the rules are loaded into the knowledge-base (by the number of terms in the left hand sides of the rules—notice that P_{14} is added to several rules to change their distinctiveness level). The priority of rules is defined to be 0 for rules 1 and 9, 2 for rules 5 and 7, and 1 for other rules. Rules 1 and 9 have the lowest priority, because we will fire rule 2 instead of rule 1 if both are enabled at the same time, and fire rule 10 instead of rule 9 if both are enabled at the same time. Rules 5 and 7 have the highest priority because these rules can be fired again only after the firings of rules 4 and 6, respectively.

The resulting ECS for the FMS can be modeled with S (as it is explained in Section 3) and we will assume that the initial states are $x_0^s \in \mathcal{X}_0^s \subset \mathcal{X}^s$ and

$$\mathcal{X}_0^s = \left\{ x^s \in \mathcal{X}^s : \sum_{i=1}^{6} \frac{x_i}{6} \in \mathbb{N}, \mathbf{x}^b = 0 \right\}. \tag{6.15}$$

Analysis It is obvious that the open-loop plant has cyclic properties. For example, when the initial open-loop plant state is $[2, 2, 3, 3, 1, 1]^\top$, then for an event trajectory "$m_{35}m_{56}m_{64}m_{43}$" the state trajectory will return to this state. The cyclic properties may prevent the open-loop plant from achieving the desired control objective (balanced part distribution). Furthermore, the part distribution in the open-loop plant may become less balanced, due to certain event trajectories. For instance, if the initial state is $[2,2,3,3,1,1]^\top$, then for an event trajectory "$m_{13}m_{21}m_{13}m_{43}m_{43}$" the state trajectory ends up at $[1,1,8,0,1,1]^\top$. However, the open-loop plant is bounded (and stable in the sense of Lyapunov), since we assume that the total number of parts in the FMS is fixed and finite. In order to eliminate undesirable open-loop system properties and meet the closed-loop specifications (i.e., part balancing) the above expert controller is employed.

When closed-loop expert control is used for the FMS, the set

$$\mathcal{X}_{fms} = \left\{ x^s \in \mathcal{X}^s : x_j = \sum_{i=1}^{6} \frac{x_i}{6}, j = 1, 2, \ldots, 6 \right\} \tag{6.16}$$

can be shown to be invariant by simple analysis of the system dynamics. A search algorithm (the A* algorithm) was used to study the reachability properties of the FMS with the expert controller described above. Using this algorithm, we show that there exists at least one path from any given initial part distribution in the FMS (with set bounds on the maximum initial buffer levels) which leads to the state in $\mathcal{X}_{fms}$ representing a balanced part distribution. The results of our reachability analysis are stated in the following result.

Theorem 6.3 (Reachability for the FMS): *The FMS described above is $(x_0^s, \mathcal{X}_{fms})$-reachable for all $x_0^s \in \mathcal{X}_0^s$ since there exists a sequence of events to occur that produces a state trajectory $s \in \mathcal{X}(S, x_0^s, \mathcal{X}_{fms})$ for any $x_0^s \in \mathcal{X}_0^s$.*

This reachability result shows that the expert controller can make appropriate "chains of inference" to reason about how to control the FMS. We have, however, not yet shown that the control objectives related to achieving a balanced part distribution are achieved. This is addressed next.

Let the distance between two states x^s and $x^{s'}$ be defined as

$$\rho(x^s, x^{s'}) = \max \left\{ \max_{i=1,2,\ldots,6} \{|x_i - x_i'|\}, \max_{i=1,2,\ldots,11} \{|x_i^b - x_i^{b'}|\} \right\}, \tag{6.17}$$

where $x^s = (\boldsymbol{x}, \boldsymbol{x}^b)$ and $x^{s'} = (\boldsymbol{x}', \boldsymbol{x}^{b'})$.

Theorem 6.4 (Region of Asymptotic Stability for the FMS): *The invariant set $\mathcal{X}_{fms}$ for the FMS above has a region of asymptotic stability w.r.t. $\mathbf{E}_v^s$ of $\mathcal{X}_0^s$.*

Proof : Let $\sum_{i=1}^{6} x_i/6 = T$. Notice that, from Equation (6.17) and the definition of $\mathcal{X}_{fms}$,

$$\rho(x^s, \mathcal{X}_{fms}) = \max_{i=1,2,\ldots,6} \{|x_i - \bar{x}_i| : \bar{x}_i = T\}. \tag{6.18}$$

Choose $V(x^s) = \rho(x^s, \mathcal{X}_{fms})$ so that choosing $\psi_1(x) = \psi_2(x) = x$ we get $\rho(x^s, \mathcal{X}_{fms}) \leq V(x^s) \leq \rho(x^s, \mathcal{X}_{fms})$, which results in the satisfaction of condition (i) of Theorem 3.1. We want to show that if the initial state $x_0^s \in \mathcal{X}_0^s$, $V(x_k^s)$ is a nonincreasing function in k and $V(x_k^s) \to 0$ as $k \to \infty$ for the rules that the expert controller can fire.

Assume that the initial state $x_0^s \in \mathcal{X}_0^s$. Note that, for the chosen rule-base, when m_{ij} occurs, the next state has

$$x_i' = \begin{cases} x_i - \left(\frac{x_i - x_j}{2}\right) = \frac{x_i}{2} + \frac{x_j}{2} & \text{if } \frac{x_i - x_j}{2} \in \mathbb{N} \\ x_i - \left(\frac{x_i - x_j}{2} + \frac{1}{2}\right) = \frac{x_i}{2} + \frac{x_j}{2} - \frac{1}{2} & \text{if } \frac{x_i - x_j}{2} + \frac{1}{2} \in \mathbb{N} \end{cases} \tag{6.19}$$

$$x_j' = \begin{cases} x_j + \left(\frac{x_i - x_j}{2}\right) = \frac{x_i}{2} + \frac{x_j}{2} & \text{if } \frac{x_i - x_j}{2} \in \mathbb{N} \\ x_j + \left(\frac{x_i - x_j}{2} + \frac{1}{2}\right) = \frac{x_i}{2} + \frac{x_j}{2} + \frac{1}{2} & \text{if } \frac{x_i - x_j}{2} + \frac{1}{2} \in \mathbb{N} \end{cases} \tag{6.20}$$

and $x_p' = x_p$ for $p \neq i, j$. The controller chooses to fire a rule which moves parts from a machine with the maximum number of parts to its neighboring machine. Let $i^* \in \{i : x_i \geq x_j, j \in M\}$ and $j^* \in \{j : x_j \leq x_i, i \in M\}$. With the occurrence of an event "m_{i^*j}" at time k (denoted by $m_{i^*j_k}$), there are two possible cases:

Case 1: $x_{i_k^*} - x_{j_k} = \alpha > 1$: The next state has

$$\begin{aligned} x_{i_{k+1}^*} &\leq x_{i_k^*} - \frac{\alpha}{2} < x_{i_k^*} \\ x_{j_k} < x_{j_{k+1}} &\leq x_{j_k} + \frac{\alpha}{2} + \frac{1}{2} < x_{j_k} + \alpha \end{aligned}$$

and $x_{p_{k+1}} = x_{p_k}$ for $p \neq i^*, j$. Thus,

$$\max_{i=1,2,\ldots,6} \{x_{i_{k+1}}\} \leq x_{i_k^*} \text{ and } \min_{j=1,2,\ldots,6} \{x_{j_{k+1}}\} \geq x_{j_k^*} \tag{6.21}$$

Hence, it can be concluded that $V(x_k^s)$ is a nonincreasing function in k. The controller will repeatedly fire rules which transfers parts from the machine with the maximum number of parts to the one which has fewer parts. Since $x_{i_{k+1}^*} \leq x_{i_k^*} - \frac{\alpha}{2} < x_{i_k^*}$ and there are finite number of machines in the system, eventually $\max_{i=1,2,\ldots,6}\{x_{i_{k+1}}\} < x_{i_k^*}$. Furthermore, since there are finite number of machines in the system, "m_{ij^*}" will eventually occur (at a time denoted by time k'), causing $\min_{j=1,2,\ldots,6}\{x_{j_{k'+1}^*}\} > x_{j_{k'}^*}$. Hence, since $\max_{i=1,2,\ldots,6}\{x_{i_k}\}$ and $\min_{j=1,2,\ldots,6}\{x_{j_k}\}$ will eventually decrease and increase with k, respectively, it can be concluded that $V(x_k^s)$ is a nonincreasing function in k and $V(x_k^s) \to 0$ as $k \to \infty$.

Case 2: $x_{i_k^*} - x_{j_k} = 1$: The next state has

$$\begin{aligned} x_{i_{k+1}^*} &= x_{i_k^*} - 1 \\ x_{j_{k+1}} &= x_{j_k} + 1. \end{aligned}$$

Thus, $\max_{i=1,2,\ldots,6}\{x_{i_{k+1}}\} = x_{i_k^*} = x_{j_{k+1}}$. Hence, if there exists k' and j, such that for all $k'' \geq k'$, $x_{i_{k''}^*} - x_{j_{k''}} = 1$ we will not get a balanced load. If such a k' does not exist then it must be the case that Case 1 above occurs until $V(x_k) = 0$. We can only get a persistent imbalance of 1 by the existence of cycles in the occurrence of events; however, our rule-base does not admit such behavior: Note that, from the rule-base that firing rule 5, which enables m_{34} disables rule 5 until the firing of rule 4, which enables m_{35}. Firing rule 7, which enables m_{43} disables rule 7 until the firing of rule 6, which enables m_{42}. This is to avoid transferring parts in the upper loop machines only (machines 1, 2, 3, 4) or in the lower loop machines only (machines 3, 4, 5, 6). The expert controller prohibits the occurrences of plant command input events m_{12} and m_{21} (or vice versa) or m_{56} and m_{65} (or vice versa) in consecutive orders to avoid passing parts back and forth between two machines. Hence, it must be the case that within a finite number of steps, we will either return to Case 1 or end up with $V(x_k) = 0$. ∎

Finally, we note that one could have also used a search algorithm to verify that $x_0^s \in \mathcal{X}_0^s$ converges to the invariant set $\mathcal{X}_{fms}$. This is not necessary in this case, however, as it is easy to pick an appropriate Lyapunov function to verify the stability properties.

Discussion: Simulation of the ECS The sequence of events executed, when the initial number of parts in machine 5 is 12 parts and there are no parts in the other machines initially, was (m_{ij_k} means transfer parts from i to j at time k): m_{00_0}, m_{56_1}, m_{64_2}, m_{56_3}, m_{64_4}, m_{43_5}, m_{64_6}, m_{56_7}, m_{64_8}, m_{42_9}, $m_{64_{10}}$, $m_{43_{11}}$, $m_{34_{12}}$, $m_{42_{13}}$, $m_{21_{14}}$, $m_{56_{15}}$, $m_{64_{16}}$, $m_{43_{17}}$, $m_{35_{18}}$, $m_{56_{19}}$, $m_{64_{20}}$, $m_{42_{21}}$, $m_{00_{22}}$, $m_{00_{23}}$, $m_{00_{24}}$, and $m_{00_{25}}$. The number of parts in each machine at times $k = 15, 16, 17$ is the same as at times $k = 19, 20, 21$. At time $k = 15$, we have a maximum imbalance of 1 (machine 5 has 3 parts, machine 2 has 1 part, and all the others have 2 parts), so the expert controller tries to balance the part distribution by enabling a sequence of event trajectories, which will eventually result in reducing the imbalance (at time $k = 21$ in this problem). In particular, at time $k = 15$ the extra part in machine 5 is passed to machine 6. Then at time $k = 16$ the extra part is passed to machine 4. Since rule 7 has higher priority than rule 6, "m_{43}" is enabled instead of "m_{42}" at time $k = 17$. The "m_{43}" occurs at time $k = 17$, but does not balance the part distribution. At time $k = 18$, only "m_{35}" is enabled. The event "m_{34}" is disabled at time $k = 18$, since it already occurred at time $k = 12$. As a result, the part distribution at time $k = 19$ is the same as the one at time $k = 15$. After passing the parts at times $k = 19, 20$, the controller does not repeat enabling "m_{43}" at time $k = 21$ since it has been disabled since after its occurrence at time $k = 17$, and chooses "m_{42}" instead, which balances the parts. This example illustrates Case 2 of the proof of Theorem 6.4, above.

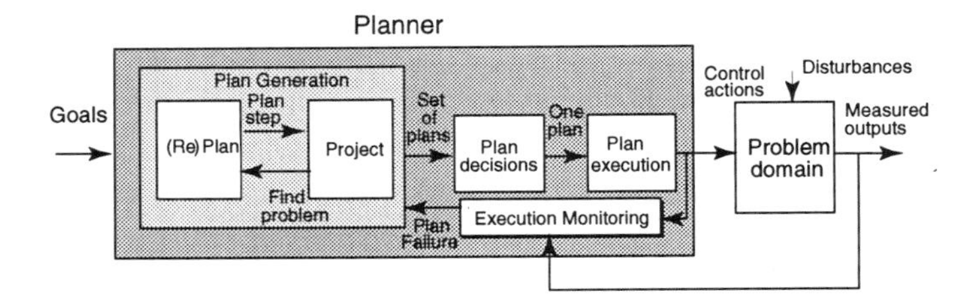

Fig. 6.5 Closed-loop planning system.

6.3 PLANNING SYSTEMS FOR CONTROL

In this section we explain how to view a planner as a controller and a planning system coupled with its problem domain as a closed-loop control system. Then, we give ideas on how to characterize and analyze controllability and stability of such systems.

6.3.1 Planners as Controllers

Artificially intelligent planning systems (computer programs that emulate the way that experts plan) have been used in path planning and high-level decisions about control tasks for robots. A generic planning system can be configured in the architecture of a standard control system, as is shown in Figure 6.5. Here, the "problem domain" (the plant) is the environment that the planner operates in. There are measured outputs y_k at step k (variables of the problem domain that can be sensed in real-time), control actions u_k, (the ways in which we can affect the problem domain), disturbances d_k (that represent random events that can affect the problem domain and, hence, the measured variable y_k), and goals g_k (what we would like to achieve in the problem domain). There are closed-loop specifications that quantify performance specifications and stability requirements.

It is the task of the planner in Figure 6.5 to monitor the measured outputs and goals and generate control actions that will counteract the effects of the disturbances and result in the goals and the closed-loop specifications to be achieved. To do this, the planner performs "plan generation," where it projects into the future (usually a finite number of steps and often a model of the problem domain is used) and tries to determine a set of candidate plans. Next, this set of plans is pruned to one plan that is the best one to apply at the current time (where "best" can be determined based on consumption of resources). The plan is executed and during execution the performance resulting from the plan is monitored and evaluated. Often, due to disturbances, plans will fail and, hence, the planner must generate a new set of candidate plans, select one, then execute that one. The "situation assessment" module tries to estimate the state of the problem domain (this can be useful in execution monitoring and in plan generation).

While not pictured in Figure 6.5, some planning systems use "world modeling," where a model of the problem domain is developed in an on-line fashion (similar to on-line system identification), and "planner design" that uses information from the world modeler to tune the planner (so that it makes the right plans for the current problem domain). The reader will, perhaps, think of such a planning system as a general adaptive controller.

6.3.2 Analysis of Closed-Loop Planning Systems

In this section we will briefly discuss analysis issues in control systems that have a planner as a controller. We will use a simple robot system to illustrate the ideas. In our robotic system we assume that we have a robot arm and manipulator that is quite dextrous. We assume that one can send the command "pick up object," and it will know how to move to the object, grip it, and pick it up. For our discussion below we use the symbol u_k to represent the input to the problem domain at time k. The symbols are quite general and allow for the representation of all possible actions that any planner can take on the problem domain. For example, $u_1 =$"pick up object," or $u_2 =$"move manipulator from position 3 to position 7." We will use y_k to represent the measurable outputs of the robotic system. For example, in the robot problem the position of some of the objects to be moved could be represented with y_k. The outputs could be $y_1 =$"object 1 in position 5" and $y_2 =$"object 1 in position 3." The inputs u_k can affect the physical system so that the outputs y_k can change over time. In our robot problem domain, the initial state can be the initial positions of the manipulator and objects. For two objects, the initial state might be $x_0 =$"object 1 in position 3 and object 2 in position 7 and manipulator in position 5." In our robotics problem domain a disturbance might be some external, unmodeled agent, who also moves the objects. We will denote the disturbance with d_k.

Controllability In control theory, and thus in planning theory, controllability refers to the ability of a system's inputs to change the state of the system. A sequence of inputs u_k can transfer or steer a state from one value to another. In the robot example, a sequence of input actions may transfer the state from $x_0 =$"object 1 in position 3 and object 2 in position 7 and manipulator in position 5" to x_7="object 1 in position 5 and object 2 in position 10 and manipulator in position 1." If a problem domain is "completely controllable," then for any state there exists a planner that can achieve any specified goal state. Sometimes complete controllability is not a property of the system, but it may possess a weaker form of controllability where the state can only be moved to certain values. For the robot problem if the problem domain is completely controllable then there exists a way for the robot to move the objects into any combination of positions.

Stability In control, and thus in planning theory, we say that a system is stable if with no inputs, when the system begins in some particular set of states and the state is perturbed, it will always return to that set of states. For the discussion we partition the state space into disjoint sets of "good" states, and "bad" states. Also we define the null input for all problem domains as the input that has no effect on the problem domain. Assume that the input to the system is the null input for all time. We say that a system is stable if when it begins in a good state and is perturbed into any other state it will always return to a good state. Similar to asymptotic stability.

To clarify the definition, a specific example is given. Suppose that we have the same robot manipulator described above. Suppose further that the set of positions the manipulator can be in can be broken into two sets, the good positions and the bad positions. A good position might be one in some envelope of its reach, while a bad one might be where it would be dangerously close to some human operator. If such a system were internally stable, then if when the manipulator was in the good envelope and was bumped by something, then it may become dangerously close to the human operator but it would resituate itself back into the good envelope without any external intervention.

The closed-loop planning system is also amenable to stability analysis. Let the reference input be a null input. The ability of the planner to maintain stable operation is clearly a fundamental property of the system. For instance, in the robot example, if the problem domain is unstable, then how do we design a planner that will ensure that the robot takes the appropriate actions so that the closed-loop system ensures that the operator is safe.

6.4 INTELLIGENT AND AUTONOMOUS CONTROL

Autonomous systems have the capability to independently perform complex tasks with a high degree of success. Consumer and governmental demands for such systems are frequently forcing engineers to push many functions normally performed by humans into machines. For instance, in the emerging area of intelligent vehicle highway systems (IVHS) engineers are designing vehicles and highways that can fully automate vehicle route selection, steering, braking, and throttle control to reduce congestion and improve safety. In manufacturing systems, efficiency optimization and flow control are being automated and robots are replacing humans in performing relatively complex tasks.

From a broad historical perspective each of these applications began at a low level of automation and through the years has evolved into a higher autonomy system. For example, today's automotive cruise controllers are the ancestors of the controllers that achieve coordinated control of steering, braking, and throttle for autonomous vehicle driving. The general trend has been for engineers incrementally to "add more intelligence" in response to consumer, industrial, and government demands and thereby create systems with increased levels of autonomy.

In this process of enhancing autonomy by adding intelligence, engineers often study how humans solve problems, then try to directly automate their knowledge and techniques to achieve high levels of automation. Other times, engineers study how intelligent biological systems perform complex tasks then seek to automate "nature's approach" in a computer algorithm or circuit implementation to solve a practical technological problem (e.g., in certain vision systems). Such approaches where we seek to emulate the functionality of an intelligent biological system (e.g., the human) to solve a technological problem can be collectively named "intelligent systems and control techniques." It is by using such techniques that some engineers are trying to create highly autonomous systems such as those listed above.

What is "Intelligent Control"? Since the answer to this question can get rather philosophical, let us focus on a working definition that does not dwell on definitions of "intelligence" (since there is no widely accepted one, since it has so many dimensions) and issues of whether we truly model or emulate intelligence but instead focuses on: (i) the methodologies used in the construction of controllers and (ii) the ability of an artificial system to perform activities normally performed by humans.

"Intelligent control" techniques offer alternative approaches to conventional approaches by borrowing ideas from intelligent biological systems. Such ideas can either come from humans who are, for example, experts at manually solving the control problem, or by observing the way in which a biological system operates and using analogous techniques in the solution of control problems. For instance, we may ask a human driver to provide a detailed explanation of how they manually solve the intervehicle distance control problem then use this knowledge directly in an expert controller. Such intelligent control techniques may exploit the information represented in a mathematical model or may heavily rely on heuristics on how to best control the process. The primary difference from conventional approaches (such as proportional-integral-derivative (PID) control) is that intelligent control techniques are motivated by the functionality of intelligent biological systems, either in how they perform the control task, or in how they provide an innovative solution to another problem that can be adapted to solve a control problem. This is not to say that systems that are not developed using intelligent systems and control techniques such as those listed above cannot be called "intelligent"; traditionally we have often called any system intelligent if it is designed to perform a task that has normally been performed by humans (e.g., we use the term "intelligent" vehicle highway systems). A full discussion on defining intelligent control involves considering additional issues in psychology, human cognition, artificial intelligence, and control. The interested reader is referred to the articles listed at the end of this chapter for a more detailed exposition that considers these issues.

There are a wide variety of techniques that can be used for intelligent control including fuzzy systems, expert systems, planning systems, genetic algorithms, and neural networks. Such approaches form "building blocks"

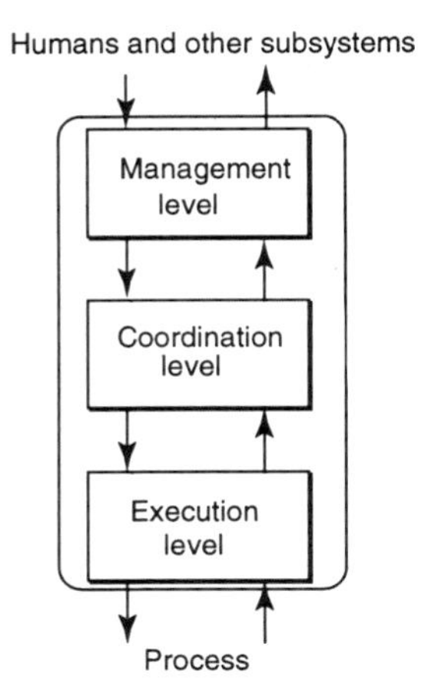

Fig. 6.6 Intelligent autonomous controller.

for intelligent autonomous control systems. Figure 6.6 shows a functional architecture for an intelligent autonomous controller with an interface to the process involving sensing (e.g., via conventional sensing technology, vision, touch, smell, etc.), actuation (e.g., via hydraulics, robotics, motors, etc.), and an interface to humans (e.g., a driver, pilot, crew, etc.) and other systems.

The "execution level" has low level numeric signal processing and control algorithms (e.g., PID, optimal, adaptive, or intelligent control; parameter estimators, failure detection and identification (FDI) algorithms). The "coordination level" provides for tuning, scheduling, supervision, and redesign of the execution level algorithms, crisis management, planning and learning capabilities for the coordination of execution level tasks, and higher-level symbolic decision-making for FDI and control algorithm management. The "management level" provides for the supervision of lower level functions and for managing the interface to the human(s) and other systems. In particular, the management level will interact with the users in generating goals for the controller and in assessing capabilities of the system. The management level also monitors performance of the lower level systems, plans activities at the highest level (and in cooperation with humans), and performs high level learning about the user and the lower-level algorithms.

Intelligent systems or intelligent controllers (e.g., fuzzy, neural, genetic, expert, etc.) can be employed as appropriate in the implementation of various functions at the three levels of the intelligent autonomous controller. For example, adaptive fuzzy control may be used at the execution level for adaptation, genetic algorithms may be used in the coordination level to pick an optimal coordination strategy, and planning systems may be used at the management level for sequencing operations. Hierarchical controllers, composed of a hybrid mix of intelligent and conventional systems, are commonly used in the intelligent control of complex dynamical systems. This is due to the fact that to achieve high levels of autonomy, we often need high levels of intelligence, which calls for incorporation of a diversity of decision-making approaches for complex dynamical learning and reasoning.

There are several fundamental characteristics that have been identified for intelligent autonomous control systems. For example, there is generally a successive delegation of duties from the higher to lower levels and the number of distinct tasks typically increases as we go down the hierarchy. Higher levels are often concerned with slower aspects of the system's behavior and with its larger portions, or broader aspects. There is then a smaller contextual horizon at lower levels, that is, the control decisions are made by considering less information. Higher levels are typically concerned with longer time horizons than lower levels. It is said that there is "increasing intelligence with decreasing precision as one moves from the lower to the higher levels" (see [93]). At the higher levels there is typically a decrease in time scale density, a decrease in bandwidth or system rate, and a decrease in the decision (control action) rate. In addition, there is typically a decrease in granularity of models used, or, equivalently, an increase in model abstractness at the higher levels.

Finally, we note that there is an ongoing evolution of the intelligent functions of an autonomous controller so that by the time one implements its functions, they no longer appear intelligent—just algorithmic. It is this evolution principle, doubts about our ability to implement "artificial intelligence," and the fact that implemented intelligent controllers are nonlinear controllers that many researchers feel more comfortable focusing on enhancing autonomy, rather than achieving intelligent behavior.

Next, it is explained how to incorporate the notion of autonomy into the conventional manner of thinking about control problems. Consider the general control system shown in Figure 6.7, where P is a model of the plant, C represents the controller, and T represents specifications on how we would like the closed loop system to behave (i.e., closed-loop specifications). For some classical control problems the scope is limited so that C and P are linear and T simply represents, for example, stability, rise time, and overshoot specifications. In this case, intelligent control techniques may not be needed. As engineers, the simplest solution that works is the best one. We tend to need more complex controllers for more complex plants (where, for example, there is a significant amount of uncertainty) and more demanding closed loop specifications T. Consider the case where:

- P is so complex that it is most convenient to represent it with ordinary differential equations and discrete event system (DES) models [36] (or some other hybrid mix of models) and for some parts of the plant the model is not known (or it is too expensive to determine); and

- T is used to characterize the desire to make the system perform well and act with high degrees of autonomy (i.e., via [3], "so that the system performs well under significant uncertainties in the system and its envi-

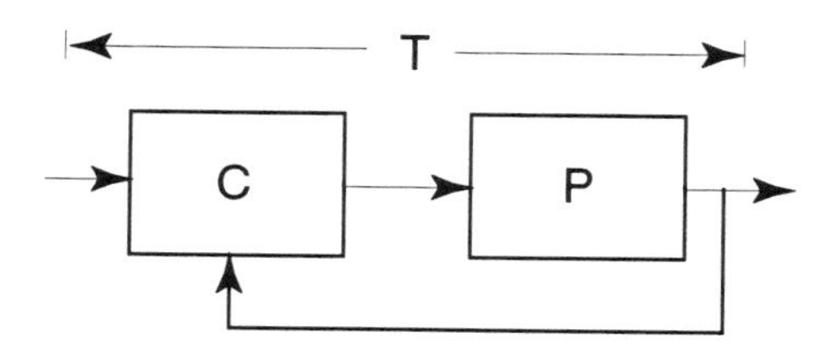

Fig. 6.7 General control system.

ronment for extended periods of time, and compensates for significant system failures without external intervention").

The general control problem is how to construct C, given P, so that T holds. The intelligent autonomous controller described briefly in the previous section provides a general architecture for C to achieve highly autonomous behavior specified by T for very complex plants P.

From a control engineer's perspective, researchers in the field of intelligent control are trying to use intelligent (and conventional) control methodologies to solve the general control problem (i.e., they are trying to find C to enhance autonomy). In reality, researchers in intelligent (and conventional) control are often examining portions of the above general control problem and trying to make incremental progress toward a solution. For example, a simple direct fuzzy controller could, perhaps, be called an "intelligent controller," but not an "autonomous controller," as most do not achieve high levels of autonomous operation, but merely help enhance performance like many conventional controllers; adaptive and supervisory approaches slightly increase performance but do not achieve full autonomy (some would never call a controller without adaptation capabilities an intelligent controller). It is important to note that researchers in intelligent control have been naturally led to focus on the very demanding general control problem described above: (1) in order to address pressing needs for practical applications, and (2) since often there is a need to focus on representing more aspects of the process, so that they can be used to reduce the uncertainty in making high-level decisions about how to perform control functions that are normally performed by humans.

Have we achieved autonomous control via intelligent control or any other methods? This is a difficult question to answer, since certain levels of autonomy have certainly been achieved but there are no rigorous definitions of "degrees of autonomy." For instance, (relatively) autonomous robots and autonomous vehicles have been implemented. It is clear that current intelligent systems only roughly model their biological counterparts and, hence, from one perspective they can achieve relatively little. What will we be able to do if we succeed in emulating their functions (and form)? Achieve full autonomy via the correct orchestration of intelligent control? Considering the benefits that will be realized on the path to this ultimate goal of achieving autonomy, perhaps it is the pursuit that matters most. The focus is certainly opening up

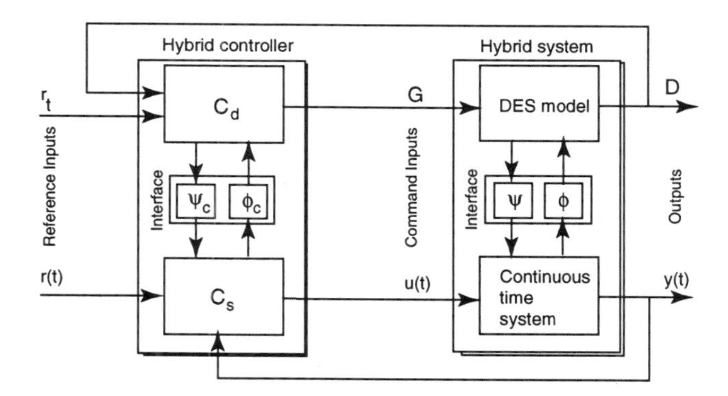

Fig. 6.8 A hybrid controller for a hybrid plant.

new horizons for the field of control, and providing exciting, promising, and productive research directions that we can get involved in.

6.5 HYBRID SYSTEM THEORY AND ANALYSIS

A hybrid system is one that has components that are easiest to model with a mix of different models. Recently, this has often come to mean a mix of conventional differential equation models with discrete event system models. For instance, a hybrid model would be one that is composed of an ordinary differential equation and a Petri net. There are many processes that are actually hybrid in nature and, hence, dictate the need for hybrid models. For example, an automated highway system has low-level steering, braking, and throttle controls and high-level controls for platoon maneuvering, check-in, and check-out procedures, and so on. Moreover, there are many times that it may be desirable to implement a hybrid controller for a system (e.g., an intelligent controller as described in the previous section) that may be easily described by conventional differential equations. In this case, the closed-loop system becomes hybrid, due to the presence of a hybrid controller. We see that a system could be hybrid, due to the plant, the controller, or both.

A hybrid control system is shown in Figure 6.8. Notice that there is a continuous time system and DES component to the plant, with an "interface." This interface has a function ϕ which maps conditions in the continuous time system to events in the DES. It also has a function ψ that maps conditions in the DES to changes to the continuous time system. The controller has two components. There is C_d, the controller for the discrete event component of the plant, and C_s, the controller for the continuous time component of the plant (this is only one way to set up the controller structure). The controller has a similar interface to the plant model and a similarly structured model could be used for the closed-loop system.

As for the other applications in this book there is often the need to conduct stability analysis of hybrid systems. It is possible to define some hybrid systems on a metric space so that the approaches of this book will apply directly. It is an important research direction to develop modeling, analysis, and design methodologies, which will be guaranteed to produce stable and high performance hybrid systems. If this problem can be solved then we will also know much more about how to conduct stability analysis for complex intelligent autonomous control systems. Some progress in modeling and stability analysis of hybrid systems is summarized in Section 6.7, "For Further Study."

6.6 SUMMARY

In this chapter we have shown how to model and analyze stability and reachability properties of expert control systems. Also, we briefly discuss the analysis of control systems that use planning systems as their controllers and general intelligent autonomous control. We close with a brief discussion on hybrid systems. The main topics covered in this chapter are:

- Expert controller and its model;
- Antecedents, consequents, and rules;
- Inference strategies, matching;
- Closed-loop expert control system;
- Meaning of reachability, cyclic, and stability properties;
- Tank and flexible manufacturing system examples;
- Planning systems as controllers and their analysis;
- Intelligent autonomous control; and
- What a hybrid system is, and why analysis of stability properties is important for hybrid systems.

The last three topics were only covered in a cursory fashion, and each of these presents opportunities for further research.

6.7 FOR FURTHER STUDY

For more information on "rule-based" expert systems see [11, 10, 28, 103, 3]. Other studies in expert system verification are given in [34, 11, 28, 103, 31, 33, 20, 39, 30, 65, 95, 63, 62, 64, 87]. The expert system section of this chapter is based on the work in [49, 76, 50]. For the expert controller, the knowledge-base and inference mechanism are modeled here, using an approach similar to that used in [49]. Our approach to modeling the rule-base is most similar to the work in [32] where the authors show how to model rule-based AI systems with a high-level Petri net [61]. In comparison with other related work, the emphasis on the development of the ECS in this chapter is different from the work done by Åström et al. in [4, 5], where the expert controller was first introduced. The primary difference is that the expert controller in this framework uses the knowledge-base, inference engine, plant output, and reference input to *directly* produce the command input to the plant. The work in [4, 5] uses the rule-based expert system for selecting/tuning conventional control algorithms (i.e., in a hierarchical, supervisory control mode).

In [50] a single model that can represent both the knowledge-base and inference engine *and* an interface to user inputs and process data and information is introduced. Using this, the authors show how to analyze qualitative properties (including reachability, cyclic behavior, and stability) of the isolated expert controller. They provide a detailed comparison between the analysis of such qualitative properties and the analysis of *static properties* ("consistency" and "completeness") of knowledge-bases. Moreover, in [50], the authors verify the qualitative properties of an expert system that solves a "water-jug filling problem," where there is no interface with users or a dynamic process, and a simple process control problem, where user inputs and dynamic process data are used in the inference process of the expert system. For more information on planning systems, see [70], and for intelligent control systems, see [3].

Early work on modeling and analysis of hybrid systems was done in [83, 84] (and this work influences our discussions in this chapter, particularly in how we structure Figure 6.8), and some work that later discussed reachability and stability analysis is in [74]. For more information on hybrid systems, see [2]. Some recent work on the area of stability analysis of hybrid systems is given in [25, 8, 26, 7, 27, 105, 104, 51, 106] and stability and other issues are discussed in the April 1998 *IEEE Transactions on Automatic Control* special issue on hybrid systems.

References

1. Anthony Angsana and Kevin M. Passino. Distributed fuzzy control of flexible manufacturing systems. *IEEE Transactions on Control Systems Technology*, 2:423–435, December 1994.

2. P. Antsaklis, W. Kohn, A. Nerode, and S. Sastry, editors. *Hybrid Systems II*. Lecture Notes in Computer Science, LNCS 999, Springer-Verlag, NY, 1995.

3. P. J. Antsaklis and K. M. Passino, editors. *An Introduction to Intelligent and Autonomous Control*. Kluwer Academic Publishers, Norwell, MA, 1993.

4. K. J. Åström, J. J. Anton, and K. E. Årzén. Expert control. *Automatica*, 22:277–286, 1986.

5. K. J. Åström and K. E. Årzén. Expert control. In P. J. Antsaklis and K. M. Passino, editors, *An Introduction to Intelligent and Autonomous Control*. Kluwer Academic Publishers, Norwell, MA, 1993.

6. R.K. Boel and J.H. van Schuppen. Distributed routing for load balancing. In *Proceedings of the IEEE*, pages 210–221, January 1989.

7. Michael S. Branicky. Analysis of continuous switching systems: Theory and examples. In *Proc. American Control Conference*, pages 3110–3114, Baltimore, MD, June 1994.

8. Michael S. Branicky. Stability of switched and hybrid systems. In *Proc. 33rd IEEE Conf. Decision Control*, pages 3498–3503, Lake Buena Vista, FL, Dec. 1994.

9. Y. Brave and M. Heymann. On stabilization of discrete event processes. In *Proceeding of the 28th Conf. on Dec. and Control*, pages 2737–2742, Tampa, FL, Dec. 1989.

10. L. Brownston, R. Farrell, E. Kant, and N. Martin. *Programming Expert Systems in OPS5*. Addison Wesley, Reading, MA, 1986.

11. B. Buchanan and E. H. Shortliffe. *Rule Based Expert Systems, MYCIN*. Addison Wesley, Reading, MA, 1984.

12. J.R. Buchi. On a decision method in restricted second order arithmetic. In *Proc. of the Int. Congress on Logic, Mathematics, and Phil. of Sci., 1960*, pages 1–11, Stanford, CA, 1962. Stanford Univ. Press.

13. Kevin Burgess. Stablility and boundedness analysis of discrete event systems. Master's thesis, The Ohio State University, Department of Electrical Engineering, May 1992.

14. Kevin Burgess. *Stability and Performance Analysis of Scheduling Policies for Flexible Manufacturing Systems*. PhD thesis, The Ohio State University, Department of Electrical Engineering, March 1995.

15. Kevin Burgess and Kevin M. Passino. Stability analysis of load balancing systems. In *Proceedings of American Control Conference*, pages 2415–2419, San Francisco, CA, June 1993.

16. Kevin Burgess and Kevin M. Passino. Stable scheduling policies for flexible manufacturing systems. *IEEE Trans. on Automatic Control*, 42(3):420–425, 1997.

17. Kevin Burgess and Kevin M. Passino. Path clearing policies for flexible manufacturing systems. *To appear in the IEEE Trans. on Automatic Control*, 1998.

18. K.L. Burgess and K.M. Passino. Stability analysis of load balancing systems. *Int. Journal of Control*, 61(2):357–393, Feb. 1994.

19. Christos G. Cassandras. *Discrete Event Systems*. Richard D. Irwin, Inc. and Aksen Associates, Inc., 1993.

20. B. Chandrasekaran. On evaluating AI systems for medical diagnosis. *The AI Magazine*, pages 34–48, Summer 1983.

21. Christopher Chase and Peter J. Ramadge. On real-time scheduling policies for flexible manufacturing systems. *IEEE Transactions on Automatic Control*, 37:491–496, April 1992.

22. E.M. Clarke, M.C. Browne, E.A. Emerson, and A.P. Sistla. Using temporal logic for automatic verification of finite state systems. In K.R. Apt, editor, *Logics and Models of Concurrent Systems*, pages 3–25. Springer Verlag, NY, 1985.

23. Rene L. Cruz. A calculus for network delay, Part I: Network elements in isolation. *IEEE Transactions on Information Theory*, 37:114–131, January 1991.

24. E.W. Dijkstra. Self-stabilizing systems in spite of distributed control. In *Comm. of the ACM*, pages 643–644, November 1974.

25. M. Dogruel and Ü. Özgüner. Controllability, reachability, stabilizability and state reduction in automata. In *IEEE International Symposium on Intelligent Control*, Glasgow, Scotland, 1992.

26. M. Dogruel and Ü. Özgüner. Stability of hybrid systems. In *IEEE International Symposium on Intelligent Control*, Columbus, OH, 1994.

27. M. Dogruel and Ü. Özgüner. Modeling and stability issues in hybrid systems. In P. Antsaklis, W. Kohn, A. Nerode, and S. Sastry, editors, *Hybrid Systems II*, pages 148–165. Lecture Notes in Computer Science, LNCS 999, Springer-Verlag, NY, 1995.

28. F. Hayes-Roth, et al. *Building Expert Systems.* Addison Wesley, Reading, MA, 1985.

29. A. Fusaoka, H. Seki, and K. Takahashi. A description and reasoning of plant controllers in temporal logic. In *Proc. of the 8th Int. Joint Conf. on Artificial Intelligence*, pages 405–408, August 1983.

30. J. Gaschnig, P. Klahr, H. Pople, E. Shortliffe, and A. Terry. Evaluation of expert systems: Issues and case studies. In F. Hayes-Roth, D. A. Waterman, and D. B. Lenat, editors, *Building Expert Systems.* Addison Wesley, Reading, MA, 1983.

31. J. R. Geissman. Verification and validation of expert systems. *AI Expert*, 3(2):26–33, February 1988.

32. A. Giordana and L. Saitta. Modeling production rules by means of predicate transition networks. *Inf. Sci*, 39:1–41, 1985.

33. C. J. Green. Verification and validation of expert systems. In *Proceedings of IEEE Western Conference on Expert Systems*, pages 28–43, June 1991.

34. U.G. Gupta, editor. *Validating and Verifying Knowledge-Based Systems.* IEEE Computer Society Press, Los Alamitos, CA, 1991.

35. W. Hahn. *Stability of Motion.* Springer-Verlag, NY, 1967.

36. L. Ho, editor. *Discrete Event Dynamic Systems: Analyzing Complexity and Performance in the Modern World.* IEEE Press, NY, 1992.

37. Y.-C. Ho and X.-R. Cao. *Perturbation Analysis of Discrete Event Dynamic Systems.* Kluwer Academic Publishers, Norwell, Massachusetts, 1991.

38. Carlos Humes, Jr. A regulator stabilization technique: Kumar-Seidman revisited. *IEEE Transactions on Automatic Control*, 39:191–196, January 1994.

39. S. Kim. *Checking a Rule Base with Certainty Factor for Incompleteness and Inconsistency.* Springer Verlag, NY, 1988. In "Uncertainty and Intelligent Systems," Lecture Notes in Computer Science, Number 313.

40. J.F. Knight and K.M. Passino. Decidability for a temporal logic used in discrete event system analysis. *International Journal of Control*, 52(6):1489–1506, 1990.

41. P.R. Kumar. Re-entrant lines. *Queueing systems: theory and applications*, 13(1–3):87–110, May 1993.

42. P.R. Kumar and S.H. Lu. Distributed scheduling based on due dates and buffer priorities. *IEEE Transactions on Automatic Control*, 36:1406–1416, December 1991.

43. P.R. Kumar and Thomas J. Seidman. Dynamic instabilities and stabilization methods in distributed real-time scheduling of manufacturing systems. *IEEE Transactions on Automatic Control*, 35:289–298, March 1990.

44. Ratnesh Kumar and Vijay K. Garg. *Modeling and Control of Logical Discrete Event Systems.* Kluwer Academic Pub., Norwell, MA, 1995.

45. W. Levine, editor. *The Control Handbook.* CRC Press, Boca Raton, FL, 1996.

46. Y. Li. *Control of Vector Discrete Event Systems.* PhD thesis, University of Toronto, May 1991.

47. Y. Li and W.M. Wonham. Linear integer programming techniques in the control of vector discrete-event systems. In *Proc. of the 27th Allerton Conf. on Comm., Control, and Computing*, pages 528–537, Univ. of Illinois at Champaign–Urbana, September 1989.

48. Sheldon Lou, Suresh Sethi, and Gerhard Sorger. Analysis of a class of real-time multiproduct lot scheduling policies. *IEEE Transactions on Automatic Control*, 36:243–248, February 1991.

49. A. D. Lunardhi and K. M. Passino. Verification of dynamic properties of rule-based expert systems. In *Proceedings of the IEEE Conference on Decision and Control*, pages 1561–1566, Brighton, UK, December 1991.

50. A.D. Lunardhi and K. M. Passino. Verification of qualitative properties of rule-based expert systems. *International Journal of Applied Artificial Intelligence*, 9(6):587–621, Nov./Dec. 1995.

51. Ü. Özgüner M. Dogruel and S. Drakunov. Sliding mode control in discrete state and hybrid systems. *IEEE Transactions on Automatic Control*, 41(3):414–419, 1996.

52. Z. Manna and A. Pnueli. Verification of concurrent programs: A temporal proof system. *Dept. of Computer Science, Stanford Univ., Report No. STAN-CS-83-967*, 1983.

53. A. N. Michel and R.K. Miller. On stability preserving mappings. *IEEE Transactions on Circuits and Systems*, 30(9):671–679, September 1983.

54. A.N. Michel. Quantitative analysis of simple and interconnected systems: Stability, boundedness and trajectory behavior. *IEEE Trans. on Circuit Theory*, 17:292–301, 1970.

55. A.N. Michel and R.K. Miller. *Qualitative Analysis of Large Scale Dynamical Systems.* Academic Press, NY, 1977.

56. A.N. Michel and R.K. Miller. *Ordinary Differential Equations.* Academic Press, NY, 1982.

57. A.N. Michel and D.W. Porter. Practical stability and finite-time stability of discontinuous systems. *IEEE Trans. on Circuit Theory*, CT-19, No. 2:123–129, 1972.

58. A.N. Michel, K. Wang, and K. M. Passino. Comparison theory for general motions of dynamical systems with applications to discrete event systems. In *Proc. Int. Symp. on Circuits and Systems, San Diego, CA*, May 1992.

59. A.N. Michel, K. Wang, and K.M. Passino. Qualitative equivalence of dynamical systems with applications to discrete event systems. In *Proc. of the IEEE Conf. on Decision and Control*, pages 731–736, Tucson AZ, December 1992.

60. D.E. Muller. Infinite sequences and finite machines. In *Proc. Fourth Annual IEEE Symp. Switching Circuit Theory and Logic Design*, pages 3–16, Chicago, IL, 1963.

61. T. Murata. Petri nets: Properties, analysis, and applications. In *Proc. of the IEEE*, pages 541–580, April 1989.

62. T. A. Nguyen. Verifying consistency of production systems. In *Proceedings or the 3rd Conference on AI Applications*, FL, 1987.

63. T. A. Nguyen, W. A. Perkins, T. J. Laffey, and D. Pecora. Checking an expert systems knowledge base for consistency and completeness. *Proceedings of the 9th International Joint Conference on Artificial Intelligence*, 1:375–378, August 1985.

64. T. A. Nguyen, W. A. Perkins, T. J. Laffey, and D. Pecora. Knowledge base verification. *The AI Magazine*, pages 69–75, Summer 1987.

65. R. M. O'Keffe, O. Balci, and E. P. Smith. Validating expert system performance. *IEEE Expert*, 2(4):81–89, 1987.

66. J. Olsder and J.-P. Quadrat. *Synchronization and Linearity: An Algebra for Discrete Event Systems.* John Wiley and Sons, Chichester, England, 1992.

67. C.M. Ozveren. *Analysis and Control of Discrete Event Dynamic Systems: A State Space Approach.* PhD thesis, MIT, Cambridge, MA, LIDS Report LIDS-TH-1907, Aug., 1989.

68. C.M. Ozveren, A.S. Willsky, and P.J. Antsaklis. Stability and stabilization of discrete event dynamic systems. *Journal of the Association of Computing Machinery*, 38(3):730–752, 1991.

69. K. M. Passino and P. J. Antsaklis. On the optimal control of discrete event systems. In *Proceedings of the Conference on Decision and Control*, pages 2713–2718, Tampa, FL, December 1989.

70. K. M. Passino and P. J. Antsaklis. A system and control theoretic perspective on artificial intelligence planning systems. *International Journal of Applied Artificial Intelligence*, 3:1–32, 1989.

71. K. M. Passino and P. J. Antsaklis. Optimal stabilization of discrete event systems. In *Proceedings of the Conference on Decision and Control*, pages 670–671, Honolulu, Hawaii, December 1990.

72. K. M. Passino and P. J. Antsaklis. *Modeling and Analysis of Artificially Intelligent Planning Systems, Chapter 8 in An Introduction to Intelligent and Autonomous Control.* Kluwer Academic Press, Norwell, MA, Massachussetts, 1993.

73. K. M. Passino and P. J. Antsaklis. A metric space approach to the specification of the heuristic function for the A* algorithm. *IEEE Transactions on Systems, Man, and Cybernetics*, 24(1):159–166, Jan. 1994.

74. K. M. Passino and Ü. Özgüner. Modeling and analysis of hybrid systems: Examples. In *IEEE International Symposium on Intelligent Control*, Arlington, VA, 1991.

75. Kevin M. Passino, Kevin Burgess, and Anthony N. Michel. Lagrange stability and boundedness of discrete event systems. *Journal of Discrete Event Dynamic Systems: Theory and Applications*, 5:383–403, 1995.

76. Kevin M. Passino and Alfonsus D. Lunardhi. Qualitative analysis of expert control systems. In M. Gupta and N. Sinha, editors, *Intelligent Control: Theory and Practice*, chapter 16, pages 404–442. IEEE Press, Piscataway, NJ, 1996.

77. Kevin M. Passino, A.N. Michel, and P.J. Antsaklis. Stability analysis of discrete event systems. In *Proc. of the 28th Allerton Conf. on Communication, Control, and Computing, University of Illinois, Urbana-Champaign*, pages 487–496, October 1990.

78. Kevin M. Passino, A.N. Michel, and P.J. Antsaklis. Lyapunov stability of a class of discrete event systems. *IEEE Transactions on Automatic Control*, 37:269–279, February 1994.

79. K.M. Passino and P.J. Antsaklis. Branching time temporal logic for discrete event system analysis. In *Proc. of the 26th Allerton Conf. on Communication, Control, and Computing, Univ. of Illinois at Urbana-Champaign*, pages 1160–1169, Sept. 1988.

80. K.M. Passino and A.N. Michel. Stability and boundedness analysis of discrete event systems. In *Proc. of the American Control Conference*, pages 3201–3205, Chicago, IL, June 1992.

81. K.M. Passino, A.N. Michel, and P.J. Antsaklis. Lyapunov stability of a class of discrete event systems. In *Proc. of the American Control Conference*, pages 2911–2916, Boston, MA, June 1991.

82. J. Pearl. *Heuristics: Intelligent Search Strategies for Computer Problem Solving.* Addison Wesley, Reading, MA, 1984.

83. P. Peleties and R. DeCarlo. Modeling of interacting continuous time and discrete event systems: An example. In *Proc. of the 26th Allerton Conf. on Communication, Control, and Computing*, pages 1150–1159, Oct. 1988.

84. P. Peleties and R. DeCarlo. A modeling strategy with event structures for hybrid systems. In *Proc. of the 28th Conf. on Decision and Control, Tampa, FL*, pages 1308–1313, Dec. 1989.

85. James R. Perkins, Carlos J. Humes, Jr., and P.R. Kumar. Distributed scheduling of flexible manufacturing systems: Stability and performance. *IEEE Transactions on Robotics and Automation*, 10(2):133–141, April 1994.

86. J.R. Perkins and P.R. Kumar. Stable, distributed, real-time scheduling of flexible manufacturing/assembly/disassembly systems. *IEEE Transaction on Automatic Control*, 34:139–148, February 1989.

87. W. A. Perkins, T. J. Laffey, D. Pecora, and T. A. Nguyen. Knowledge base verification. In G. Guida and C. Tasso, editors, *Topics in Expert System Design*. Elsevier Science Publishers B. V., North Holland, 1989.

88. J.L. Peterson. *Petri Net Theory and the Modeling of Systems*. Prentice-Hall, Engelwood Cliffs, NJ, 1981.

89. Proceedings of the IEEE. *Special Issue on Dynamics of Discrete Event Systems*, January 1989. Vol. 77, No. 1.

90. P. J. Ramadge and W. M. Wonham. Supervisory control of a class of discrete event processes. *SIAM J. Control and Optimization*, 25(1):206–230, January 1987.

91. P. J. G. Ramadge and W. M. Wonham. The control of discrete event systems. *Proceedings of the IEEE*, 77(1):81–98, January 1989.

92. M. Raynal. *Algorithms for Mutual Exclusion*. MIT Press, Cambridge, MA, 1986.

93. George N. Saridis. Analytical formulation of the principle of increasing precision with decreasing intelligence for intelligent machines. *Automatica*, 25(3):461–467, 1989.

94. Thomas I. Seidman. 'First Come, First Served' Can Be Unstable! *IEEE Transactions on Automatic Control*, pages 2166–2177, October 1994.

95. M. Suwa, A. C. Scott, and E. H. Shortliffe. An approach to verifying completeness and consistency in a rule-based expert systems. *AI Magazine*, pages 16–21, Fall 1982.

96. S. Takai, T. Ushio, and S. Kodama. Stabilization and blocking in state feedback control of discrete event systems. *Discrete Event Dynamic Systems: Theory and Applications*, 5:33–57, 1995.

97. Z.B. Tang and L.Y. Shi. Note on "Distributed scheduling based on due dates and buffer priorities" by S.H. Lu and P.R. Kumar. *IEEE Transactions on Information Theory*, 37:1661–1662, October 1992.

98. J.G. Thistle and W.M. Wonham. Control problems in a temporal logic framework. *International Journal Control*, 44(4):943–976, 1986.

99. J.N. Tsitsiklis. On the stability of asynchronous iterative processes. *Mathematical System Theory*, 20:137–153, 1987.

100. J.N. Tsitsiklis and D.P. Bertsekas. *Parallel and Distributed Computation*. Prentice-Hall, Inc., Engelwood Cliffs, NJ, 1989.

101. K. Wang, A.N. Michel, and K.M. Passino. On stability preserving mappings and qualitative equivalence of general dynamical systems: Part i: Theory. *Avomaticka i Telemekhanika (Int. Journal of Automation and Remote Control)*, 10:3–12, 1994.

102. K. Wang, A.N. Michel, and K.M. Passino. On stability preserving mappings and qualitative equivalence of general dynamical systems: Part ii: Applications. *Avomaticka i Telemekhanika (Int. Journal of Automation and Remote Control)*, 11:49–58, 1994.

103. S. M. Weiss and C. A. Kulikowski. *A Principal Guide To Designing Expert System, (Chapter 6)*. Rowman and Allenheld Pub., NJ, 1984.

104. Hui Ye, Anthony N. Michel, and Panos J. Antsaklis. A general model for the qualitative analysis of hybrid dynamical systems. In *Proc. of the 34th Conf. on Decision and Control, New Orleans*, pages 1473–1477, Dec. 1995.

105. Hui Ye, Anthony N. Michel, and Ling Hou. Stability theory for hybrid dynamical systems. In *Proc. of the 34th Conf. on Decision and Control, New Orleans*, pages 2679–2684, Dec. 1995.

106. Hui Ye, Anthony N. Michel, and Ling Hou. Stability analysis of discontinuous dynamical systems with applications. In *Proceedings of the 13th World Congress of the IFAC, Volume E, SanFrancisco, CA*, pages 461–466, June 1996.

107. T. Yoshizawa. *Stability Theory by Liapunov's Second Method.* Math. Soc. of Japan, Japan, 1966.

108. B.P. Zeigler. *Multifacetted Modeling and Discrete Event Simulation.* Academic Press, NY, 1984.

109. B.P. Zeigler. Devs representation of dynamical systems: Event based intelligent control. *Proceedings of the IEEE*, 77(1):72–80, Jan. 1989.

110. V.I. Zubov. *Methods of A.M. Lyapunov and their Application.* Noordhoff Ltd., The Netherlands, 1964.

Index

Arc weight function, 14
Arc, 14
Architecture, 185
Asymptotic stability in the large, 26
Asymptotic stability, 25, 29, 41
Asymptotically stable in the large, 30
Automata, 35
Autonomous controller, 186
Autonomous system, 182
Bounded delay, 105
Boundedness, 27
Cellular structure, 134, 138
Clear-a-fraction policy, 89
Clear-average-oldest buffer policy, 92
Computer network, 18, 43
Cyclic, 34, 164
Deadlock, 9
Demultiplexer, 99
Discrete event systems
  definition, 1
Enable function, 9
Event trajectories, 9
  allowed, 10
  valid, 9
Event
  definition, 9
  enabled, 9
  null, 9
Expert control system, 148
Expert controller, 149
Exponential stability in the large, 26
Exponential stability, 26, 30, 41
Exponentially stable in the large, 31
Feedforward line, 133
Finite state systems, 35
Finite time stability, 28
Firing count vector, 16
First-come-first-serve policy, 95
Functional architecture, 185
Global synchronous clear-a-fraction policy, 117
Global synchronous clear-largest-buffer policy, 122
Global synchronous periodic clearing policy, 123
Graph, 10, 12
Hierarchical control, 184
Hybrid system, 187
Incidence matrix, 15
Intelligent control, 147, 183
Invariant, 25
Lagrange stability, 27, 34
Load balancing, 12, 42, 45
Logical DES, 35
Logical time, 9
Lyapunov function, 28
Manufacturing system, 43, 85, 172
Marking, 14
Matrix equations, 15
Metric space, 24
Metric, 24
Motions, 10

Multiplexer, 99
Neighborhood, 25
One buffer machine, 42
Output stream, 98
Petri net, 14, 36
  extended, 14, 16
  general, 14
Place, 14
Planner design, 181
Planning system, 180
Practical stability, 28
Priority policy, 143
Problem domain, 180
Production network, 16, 43
Random part selection policy, 95
Rate synchronized manufacturing line, 38
Re-entrant line, 131
Reachability, 34, 163
Region of asymptotic stability, 26
Region of exponential stability, 26
Scheduling, 85
Single buffer machine, 11
Situation assessment, 180
Stable in the sense of Lyapunov, 25, 28, 41
State trajectories, 9
State
  definition, 9
Stream constraints, 86
Stream modifier, 100, 113
Surge tank, 166
Tokens, 14
Traffic system, 146
Transition, 14
Uniform boundedness, 27, 31
Uniform ultimate boundedness, 27, 32
Uniformly ultimately bounded, 33
Unstable, 26